The American Intellectual Elite

The American Intellectual Elite

Charles Kadushin

Little, Brown and Company — Boston — Toronto

FIRST EDITION

T 08/74

The author is grateful to the following:

Mr. Irving Howe for permission to quote from his essay "The New York Intellectuals," which originally appeared in *Commentary*, Volume 46, October 1968.

Change Magazine for permission to include parts of the article "Influential Intellectual Journals: A Very Private Club" by Julie Hover and Charles Kadushin. Copyright © 1972 by Educational Change, Volume 4, #2; March 1972.

The Public Interest for permission to quote from the article "Who Are the Elite Intellectuals?" by Charles Kadushin. Copyright © National Affairs, Inc., 1972.

Library of Congress Cataloging in Publication Data

Kadushin, Charles.
The American intellectual elite.

Bibliography: p.
1. United States--Social conditions--1960-
2. Intellectuals--United States. 3. United States--
Intellectual life. 4. Elite (Social sciences)
I. Title.
HN65.K32 301.44'5 74-5427
ISBN 0-316-478903

Published simultaneously in Canada
by Little, Brown & Company (Canada) Limited

PRINTED IN THE UNITED STATES OF AMERICA

For My Children

Preface and Acknowledgments

This book began as a technical sociological study of networks among American intellectuals who write or are written about in the leading intellectual journals and magazines. The intellectuals themselves soon took over the writing, and the book has become more concerned with intellectuals, perhaps, than with sociology. While not the kind of history a historian might write, the book as it now stands is a sociological history of leading American intellectuals of the late 1960's. The book's concern is with time and place first, and with sociological generalizations second. This shift in priorities happened partly because of the seductive qualities of the subjects of study, and partly because I wished them to be able to read this book rather than have them dismiss it as one more piece of sociological nonsense because it contained too many statistics and too many technical words. After all, the "natives" I am writing about are not semiliterate tribesmen just emerged from the Stone Age, helpless to counter the accounts of anthropologists. Whatever one writes about literary intellectuals – provided it is itself halfway literate – will be read and replied to in kind. Thus, the professional social scientist may miss some of the details of our apparatus, but those interested can always ask me directly for detailed statistics.

Another reason I wanted to tell the story first, and let the theory as well as method come second, is the plethora of so-called theorizing about intellectuals. By definition, every intellectual is what social scientists call a "participant observer" of the intellectual world. Many utilize their expertise as observers of their own world to theorize informally about intellectuals. Yet, knowledgeable as they are, they have the disadvantage of being "insiders," so that they may take for granted what others may wonder about.* Then too, by virtue of being insiders they may sense they have a grasp of the entire picture when all they see is their own corner. Their theorizing is therefore often either limited in perspective or informal, or it assumes facts which the reader may not be acquainted with. As an "outsider," I have systematically collected the accounts of the insiders and have attempted to present their collective story straightforwardly and as dispassionately as I could, for I felt that in this field there was too much theorizing with too little fact.

Nonetheless, the very contact with the world of values and politics of the intellectuals has stimulated me willy-nilly to take a position. I began the project as a liberal, but something that happened in the course of the work has pushed me further to the left, even as many intellectuals moved to the right, so that I now consider myself a democratic socialist.

As a social scientist I believe that my personal biases must be made known to the reader. Also as a social scientist, however, I would like to consider that they affect my interpretations more than they do my gathering and reporting of data.

My interest in tracing informal networks and social circles dictated the method — anonymous interviews. Relations among people cannot be traced very well from published works, and many respondents are reluctant to reveal whom they talk to about what, if that information can be attributed to them. To guarantee complete anonymity, we cannot even release the names of our 110 respondents nor even the 62 others we tried to interview but could not reach. Obviously, a certain measure of gossip value is lost in anonymous interviews. I think much is gained. Not only is information given by respondents which

* Robert K. Merton, "Insiders and Outsiders: A Chapter in the Sociology of Knowledge," *American Journal of Sociology,* 78 (July 1972):9–47.

might otherwise not be available, but social scientists are enabled to analyze the larger picture because we are not sidetracked by personality idiosyncrasies.

I also discovered that what intellectuals have put into print at any one time may well differ from the perceptions, opinions, and feelings of those intellectuals who happen not to have been published on a particular topic. Some will say this does not matter, that only the printed record counts. I disagree. To the extent that there is such a thing as an intellectual community, to the extent that intellectuals do influence one another in ways other than print, to that extent what is not printed counts too, perhaps more than what eventually reaches the public. But the reader will have to judge this for himself.

A project such as this is expensive – one reason such empirical research on intellectuals is rare. The National Science Foundation granted funds for the data gathering and for the computer analysis of networks. The Ford Foundation granted me a year's Faculty Research Fellowship to write the results in book form. The study of intellectuals is part of a larger study of American leaders, under a grant from the National Institute of Mental Health to the Columbia University Bureau of Applied Social Research, and the American study itself is part of a comparative effort to study opinionmakers in various countries, under the aegis of a committee of the International Sociological Association and the International Political Science Association. The comparative aspects of the work have been aided by grants from the Carnegie Foundation and the Ford Foundation. I am grateful to these foundations for their help. Needless to say, they have not reviewed this work, nor are they in any way responsible for its contents nor the contents of any of the other research on leaders.

This project is also the work of many people. Thomas J. Conway was associate project director throughout the data collection phase and did many of the interviews. His help and guidance through the analysis go far beyond the chapter, directly attributed to him, on the Cold War. Monique Tichy was the indispensable project administrator and also did many interviews. Deborah Benton was the project secretary and also helped with the interviews. Other interviews were by Joel Millman, Iris Posner, Becky Coleman Curtis, Edna Kamis, Kenneth Kressel, Lida Orzek, Julie Hover, Kathy Garner.

Many of the interviewers were not paid (the largesse of foundations is strictly limited) but received their recompense directly from our respondents in terms of good, interesting conversation. Julie Hover drew the sample and analyzed the data on journals. During the analysis phase I was especially helped by Herman Kane, who wrote drafts of parts of the chapters on issues and power. Stella Manne also contributed to the analysis. Richard Alba developed the computer programs which trace cliques and networks and Peter Abrams was responsible for data processing. Sandy Mayers-Chen supervised the administration of the project in its latter phases. Phyllis Sheridan, Administrative Director of the Bureau of Applied Social Research, was also very helpful. Evelyn Garfiel gave much editorial help.

Intellectual encouragement and criticism throughout were given by Allen Barton and Bogdan Denitch, my co-project directors on the larger studies of which this is a part. Parts of the manuscript, drafts, and summaries were read and criticized by Bernard Barber, Daniel Bell, Gwen Bellisfield, Joseph Bensman, Noam Chomsky, Michael Harrington, Samuel Z. Klausner, Mark F. Plattner, Basil Rauch, David Riesman, Robert Shaplen, Ivan Vallier, and Harriet Zuckerman. A draft of the entire manuscript was commented upon by Lewis Coser and Stanley Rothman. Needless to say, these readers are not in agreement among themselves and I alone am responsible for the present text.

Finally, my debt to the 110 anonymous respondents, who constitute some of America's finest minds, cannot be publicly acknowledged. Some gave as much as an entire day, stolen from their own work. I only hope that this collective portrait offers some minor recompense.

Contents

I

The Shape of Intellectual Life in America

1

Who Are the Elite American Intellectuals?

There are almost as many works about intellectuals as there are intellectuals.[1] Intellectuals are of course masters of the word and their mastery is often used to write about themselves. So it may very well be that aside from ballplayers and actresses, intellectuals are the most overpublicized people in America. In this literature there are widely differing views as to who might be said to be an intellectual, much less an elite intellectual. For the most part, works on intellectuals follow a pattern. Almost half the work is devoted to a definition of who should be included under the title intellectual. Reference is often made to the development of the term "intelligentsia" in nineteenth century Russia. Arguments are then advanced as to the viability of the term today: are all "intellectual workers," from engineers to poets, to be considered intellectuals? Once the matter of definition is settled, the characteristics of intellectuals are debated: are intellectuals today old or young; are American intellectuals all to be found in New York City or has the growth of universities dispersed them around the countryside; is there an inner

A version of this chapter appeared as "Who Are the Elite Intellectuals?" in *The Public Interest*, 29 (Fall 1972).

clique or family of intellectuals who rule the roost; is it true that most intellectuals are Jews; is there a two- or three-culture system in which scientists and perhaps social scientists are exiled to a world separate from the literary elite? After these characteristics are decided upon, the denouement consists of an attack on or defense of the role of intellectuals: are they critical enough or too critical of the "system," are they a priesthood defending current values or are they a revolutionary wave with new and better ideas and values; or perhaps the problem is that they are beside the point and focus mainly on their own irrelevant little intellectual world.

Definitions of "Intellectual"

Many of the findings and much of the controversy depends on how one goes about defining the term "intellectual," so we shall follow the formula right from the start and spend a little time on definitions. While there is considerable debate about the precise time when the term appeared and could properly be applied to a group of persons, there is little argument that the term "intelligentsia" originated in nineteenth century Russia and referred, essentially, to persons who (1) were concerned with matters of public interest; (2) felt personal responsibility for the state and solution of these problems; (3) tended to view political and social questions as moral ones; and (4) felt obligated to do something in life as well as thought.[2] The term "intellectuals" appears in the January 14, 1898, "Manifeste des Intellectuels," published in *L'Aurore* protesting the Dreyfus affair.[3] The format would appear familiar to readers of petitions on the war in Vietnam[4] and the signatories included Zola, Anatole France, Proust, Léon Blum and others we would call intellectuals. Today, Fyvel[5] has observed that the term is in more common use in the United States than in Great Britain.

The history of the term "intellectual" is interesting, but in some ways irrelevant to our interests, for the role of the intellectual has always existed quite independently of the term. Surely some nineteenth century clergymen in the United States played roles as intellectuals. Intellectuals flourished during the Enlightenment, were the hallmark of the Reformation and Renaissance, and were the pride of ancient Greece and rabbinic Judaism. Even in preliterate societies, intellectual functions were an important aspect of the medicine-man or shaman role. We are obviously dealing with a function which in different societies is designated by different words. Since our major interest is in the contemporary United States, we must devote attention to particular processes by which some persons acquire the label of elite intellectual.

Though we often speak of the elite intellectual as if he were a person, the intellectual is a social role, for nobody is wholly an intellectual.[6] Some persons spend most of their time playing this role, some spend less time, and some are not at all in the running. One should not debate whether atomic scientists are just as much intellectuals as are literary critics. Rather, one must consider why, under varying conditions, persons who play certain occupational roles also play the role of intellectual.

Social roles can be defined in terms of their position in society, the general functions they serve, the kinds of specific output they produce, and their rank in the system of stratification. If we choose to define American intellectuals in terms of position, we might say that all college graduates are intellectuals or, more restrictively, that all 400,000 full-time college professors qualify, or still more restrictively, that only the 15 or so on the masthead of *Commentary, Partisan Review,* or the *New York Review of Books* are intellectuals. If we are concerned about function we might say that intellectuals are those who submit society and its ideas to basic criticism.[7] Again, we can be more or less elitist by being more or less fussy about the sort of criticism we have in mind. Most definitions of intellectual include something about the *content* of intellectual ideas: that is, the output of intellectuals. For example, roles concerned with abstract ideas are said to be intellectual roles.[8] Most definitions of the intellectual utilize two or three of these characteristics of the role and have an elitist bias. For example, intellectuals might be those whose major occupation requires them to deal with high-quality abstract ideas.

In our search for clarity, let us begin with what seems common to various definitions of intellectuals – the nature of the output. According to both Parsons[9] and Nettl,[10] who seem otherwise to disagree, the intellectual role is primarily concerned with culture, especially with that aspect of culture concerned with symbols which give meaning to objects and actions. Those kinds of symbols which express some moral apprehension of experience and action are called by Max Kadushin "value concepts."[11] Value concepts include such terms as "rights of man," "freedom of speech," "justice," and the like. Because these symbols are defined essentially in their application rather than in any abstract formulation, any member of the society who knows the terms can manipulate them with fair ease. Precisely because value concepts can be so easily applied to a variety of situations, most societies have a relatively small set of persons, called intellectuals, who are creatively expert in finding the relationship of one value concept to another and in tracing the use and application of these concepts in a society's tradition.

The intellectual role always stresses high quality and creativity, and so is frequently confused with high intellect of any sort. The role of poet, literary critic, artist, and such require high intellect and creativity, and in addition, are roles which place high priority on the manipulation of symbols of significance.[12] While the symbols of art and literature are not necessarily value terms, they do tend to evoke an esthetic experience – and esthetics are clearly a matter of significance. Because literary and artistic roles center about significance, in most societies litterateurs and artists are considered intellectuals, and intellectuals may frequently be found pursuing literary or artistic careers; at the very least, artists and litterateurs tend to move in intellectual circles.

One interpretation of C. P. Snow's "two cultures" argument[13] is that despite the intellectual elegance of science and the high degree of creativity it demands, scientists are usually not called intellectuals. Though Snow bemoans this fact, while others insist science is indeed artistic, his argument that science is a different sort of thing from literature, art and politics is basically sound. To the degree that science is value-neutral, and to the extent that its concern is not directly with esthetics, it does not deal with meaning and significance.[14] To be sure, there are contemporary scientists — for example, Einstein, Russell, Oppenheim or Commoner — whom most persons would list as having played intellectual roles. But this is because they have, in addition to their roles as scientists, dealt with matters that have been broadly related to human affairs. To the extent that scientists participate in, or even initiate, public debates on such problems as radiation, atomic energy, or the quality of the environment, they play intellectual roles, but more about this shortly.

Value concepts can be applied in a number of widely differing areas of life, for there is no area of human interest or activity which is not governed by values. Therefore the expert on values is also a generalist. In this view no one is an intellectual who plays the role of narrow specialist exclusively. Finally, the creative generalist attempts to communicate his findings to a variety of others who he thinks will understand him. These considerations bring us to a rough definition: an elite intellectual is one who is an expert in dealing with high-quality general ideas on questions of values and esthetics and who communicates his judgments on these matters to a fairly general audience.*

* Though we have arrived at it in a different way, our definition has the most in common with Edward Shils, *The Intellectuals and the Powers* (Chicago: University of Chicago Press, 1972), p. 3:

> In every society, however, there are some persons with an unusual sensitivity to the sacred, an uncommon reflectiveness about the nature of their universe and the rules which govern their society. There is in every society a minority of persons who, more than the ordinary run of their fellow man, are inquiring and desirous of being in frequent communication with symbols which are more general than the immediate concrete situations of everyday life and remote in their reference to both time and space. In this minority, there is a need to externalize this quest in real and written discourse, in poetic or plastic expression, in historical reminiscence or writing, in ritual performance and acts of worship.

Shils also points out that intellectuals fulfill societal needs, not only their own: "There would be intellectuals in every society even if there were no intellectuals by disposition." (p. 5)

I think my use of the notion of significance more neatly summarizes the concerns of intellectuals, however. Shils also defines intellectuals much more broadly than I do, at least for my present purposes.

But who or what is to say that a person is an expert and that his general ideas about values or esthetics are of high quality?

To begin with, quality is something that can only be determined by social systems: that is, by people acting in concert. As successive waves of literary and artistic fashions suggest, one group's high quality is not necessarily that of another. A concern with quality is also intimately tied to the communication process, because only by broad communication of an idea can its worth be evaluated by others. Obviously, persons who are not experts themselves cannot evaluate the worth of expert communications. Colleague evaluation is essential in all endeavors which manipulate high level abstractions. This evaluation is elaborately institutionalized in science and the professions. Only scientists can determine whether or not another person is a scientist.[15] Only colleagues can evaluate the qualifications or the excellence of a physician or lawyer. Thus, the operational definition of "physician" is that he is a person who other physicians think is a physician. Similarly, an elite intellectual is simply a person whom other elite intellectuals believe to be an elite intellectual.

This way of identifying intellectuals is not as circular as it sounds. Each field has a regularized method for certifying its members. There are boards of physicians and Ph.D. committees of scientists. We shall argue that in contemporary America the "board" or "examining committee" that admits a person to elite intellectual status is the editorial board of the intellectual journal. A leading intellectual is simply any person who writes regularly for leading intellectual journals and/or has his books reviewed in them. While the sheer number of appearances in these journals will be seen to be strongly related to intellectual prestige, it is not the only indicator. There are some leading intellectuals who write only infrequently but who are nevertheless highly regarded by the entourage or circle which surrounds most of these intellectual journals. In fact, for many years, the circle was the key to the certification of intellectuals. Let us look more carefully into the matter of circles and networks.

The modern division of labor tends to force roles into narrowly defined occupational limits, whereas the intellectual claims to speak about matters that are meaningful and significant for a very wide range of human interests and behavior. The claim that intellectuals engage in high-quality discussion of these matters, however, also implies that some social structure must exist which separates the wheat from the chaff. The intellectual appears disorganized, if not unorganizable, mainly because his mode of organization does not follow the most common social structure in modern life: the formal organization as typified by businesses, factories and universities. Rather, the characteristic form of intellectual social organization is the social circle — a loose network of relationships which nevertheless control and direct intellectuals and intellectual expression.

A social circle has three characteristics: though it may have a central core of notable figures, it does not have formal leadership because it has no formal roles; the relations between members are not instituted, though some traditional forms may evolve; most important, though some relationships are face-to-face, others are indirect (that is, members may be held together by a chain in which A relates to B who in turn relates to C).[16] Members of a circle relate one to another because of some interest they have in common. These interests may be economic, political, cultural, or personal. The presence of a common interest distinguishes a circle from a latent network in which chains may consist of a linking quality or property which is not salient at the moment to the members of the network. This results in the "small world" phenomenon.[17] For any given purpose — say, reaching an unknown person with given attributes — it is possible to activate a latent network such that even in a mass society the average distance from any given "target" is about seven individuals. In common with all networks, however, the full extent of any circle cannot be visible to any one participant, a fact that leads to the characteristic denial on the part of many intellectuals that they belong at all to any "group."

The loose form of a social circle is well suited to the requirements of organizing intellectual life: ideas and individuals must be rated, new ideas communicated, and opinions must be formed across the boundaries of individual formal organizations and even across the boundaries of different institutional areas.

The requirement that new ideas receive some kind of hearing is met better by loose structures than by more formal ones which tend to associate ideas with entrenched social positions.

To the extent that all social institutions serve social needs, solve social problems, or produce social consequences, to that extent they have a history. Social circles of intellectuals have grown out of particular historical circumstances which have shaped their nature and structure. In particular, as western industrial society developed, the function of circles in defining who are the leading intellectuals has undergone considerable change.

In the past two hundred years, there have been three main types of nonreligious social circles among intellectuals. Some were mainly concerned with political values, some with cultural life-style values, and some with both. (The fourth logical type, concerned with neither politics nor life-styles, covers a wide range of possibilities, including religion, but is not our concern here.) The purely political circle gave rise to the classic revolutionary cabal; the purely life-style circle gave rise to the classic Bohemia; the third existed as the classical cultural-literary-political salon. The last, because it was the least specialized, had the highest prestige. By "classical" we refer, of course, to the salon of the late eighteenth and early nineteenth centuries, because the appearance of the intellectual roles as an interstitial one – that is, a role fitting between the crevices of formal structures – became most apparent only when the rest of the society became segmented as a result of the industrial revolution, and of the decline of the church and nobility as integrating factors.[18]

New styles of what we now call counterestablishment politics, which emerged in both England and France in the eighteenth century, required more flexible social forms. So did a high level of artistic production, which began to break with established institutional arrangements for the distribution, judgment and appreciation of art.

Although the several types of social circles have somewhat different characteristics, they do intersect. The Bohemian network extends into the political, and both meet with the literary-political salon. In the United States, the pre–World War I salon of Mabel Dodge is a classic example. Part of the reason for the intersection of these circles is the fact that they grew as a consequence of common factors, and as a result, each served as a source of recruitment for the other.

Perhaps the simplest reason for intersection was size. Compared to modern circles, the most salient fact about classical intellectual circles was their relatively small absolute size, and the relatively small number of such circles, each of which drew from the same pool of intellectuals. The nineteenth century did see an enormous explosion in the number of persons who could make pretense to being intellectuals. (The only serious attempt to get a count was performed for French painters by Harrison and Cynthia White, and they found the staggering figure of four thousand serious painters.)[19] And it was precisely this explosion which made circle formation so important, for the new large numbers made communications and evaluation difficult. Nevertheless, the number of persons involved, as well as the number of circles, seems minuscule when compared to the situation in the contemporary United States.

Small size had other important consequences. The length of chains was relatively short and most persons either knew each other directly or at most through one other person. For both salons and political-revolutionary circles the face-to-face character of most relationships allowed greater ease in evaluating ideas and persons. The salon could directly make reputations. The tight network of these two types of circle also fitted their rather hierarchical form of organization. Though the hierarchy was not a bureaucratic one, it was still possible to judge the relative "pecking order" within a political or cultural circle. Classic Bohemia (now noted in absentia by the saying "The Left Bank [or Greenwich Village] isn't what it used to be") did not have the same hierarchical organization nor was the network as tight. One of the differences between Bohemian life-style and the bourgeois life-style of most political and cultural circles was the Bohemian value that performances cannot and should not be evaluated – or, as we would now put it, the importance of "doing one's own thing." A paucity of rigorous evaluation makes for a looser social structure. It is for this reason that cultural and political circles have always begrudged the Bohemians much claim to intellectual status or achievement.

Small size solved one of the most important problems of a social circle: constructing some kind of framework. Because of their ephemeral nature, all social circles require some sort of peg onto which social relations are hung. There are no contracts, no traditional relationships, and no formal products to cement the structure of a social circle. Circles, therefore, always grow around some existing formal structure, organization, or charismatic personage, though formal structures also tend to form around social circles. The source of initial cohesion may be as inconsequential as the Café des Deux Magots, or as central as Madame de Staël. In a small circle the cement is less important because the web of social relations can soon replace a formal center. To the extent that classical Bohemian circles had any form at all, the main cement was geography or the location of a particular cafe or street. The classic salon was maintained by notable women. Classical political circles were frequently based on charismatic leaders. By the latter part of the nineteenth century, however, these more informal forms were generally replaced by more formal ones.

Bohemia (now sometimes known in its various transformations as "the scene" or "the underground" or "the movement") has never fully shaken off its shapelessness. It intersects interest groups or occupations in the arts, films, and music in a way not altogether different from leisure-time interest groupings of the larger middle class society; these intersecting groups have become rallying points for different Bohemian social circles. Drugs, of course, have a built-in mechanism for insuring some form of social "connection." Perhaps because, by its very nature, Bohemia has never had its own firm standards of achievement, "success" achieved through notoriety in middle class society has frequently served as a peg for circles which purport to be the "scene." Norman Mailer, Andy Warhol and Joan Baez all have circles formed around them or their cults.

The failure of the Paris Commune of 1871 marked the beginning of the end of political social circles pegged exclusively on charismatic leaders and local street orators. In the first part of the nineteenth century, political movements were more likely to grow from social circles than the reverse. After Lenin, things were different, though circles always remained attached to ideological parties. While the art of "fellow-traveling" was honed to a fine point in the 1930's, even the most organized political groups have always had a penumbra of social circle networks surrounding them. Before the American depression, however, such networks were more characteristic of Europe than of the United States. Certainly for the Fabians it is difficult to decide which was more central – the social circle or the formal political organization. The formalization over forty years of the Communist Party, however, has seen, in reaction, a return by many to much looser forms of radical organization, called the "New Left," though the sheer size of present society makes these loose forms more difficult to maintain. At present, as we shall see, the failure of the New Left has made intellectual life even more shapeless.

While the French invented the salon, the British invented its more formal modern equivalent, the literary-political journal. As described by Lewis Coser, an early example was the *Edinburgh Review,* founded in 1807 by a group of highly educated but poor lawyers who were discontented with intellectual life and their own professional opportunities. These intellectuals were linked to the Scottish Renaissance of the eighteenth century. Several knew Adam Smith, William Robertson, Robert Burns and David Hume. Most had listened to Dugald Stewart, Whig Professor of Moral Philosophy at the University of Edinburgh. Most were members of the Academy of Physics Speculative Society, "where young intellectuals in and around Edinburgh University discussed legal reform, literary subjects, and science. . . . [They] thus had occasion to meet in close personal and intellectual contact prior to the founding of their review and at the same time to renew their ties with elders who embodied the tradition they hoped to preserve in the very effort of renewing it. They had ascertained their respective opinions, debated their points of agreement and difference, slowly formed a common outlook through common intercourse in an otherwise hostile intellectual atmosphere."[20]

In later years, when the journal was well established and reached a circulation of fifteen thousand copies, each read by an estimated three people who together constituted "practically everybody among the well-to-do who had the leisure and cultivation to read at all," the review formed the superstructure around which a literary-political salon grouped itself. As Coser describes it, the situation was now somewhat different from the first stage of face-to-face salons centered about coffeehouses in eighteenth century London.

Although authors no longer necessarily had that close personal contact with one another that had characterized the coffeehouse, they grouped themselves around the *Review*, sharing in a common universe of ideas. For many, the *Review* provided not only intellectual contact but also the setting for personal meetings. The editorial office became a kind of informal forum where writers would meet one another, the editor, or upon occasion, a reader.[21]

In this description we see a model for the modern intellectual circle, founded not upon geography or even key personages, but upon a literary political journal. The journal acted to certify the leading intellectuals who edited it and wrote for it, and its influence in opinion-making was great because it was read by everybody: that is, by major opinionmakers in that society. The structure was a combination of tightness and looseness. A tight inner core shaded off into a penumbra of indirect relationships. Not all the authors met each other directly. Because of this combination of tightness and looseness, it was possible for a relatively large number of interconnected circles to exist. In the modern metropolitan society which was developing in Britain, intellectuals could at the same time feel close to a particular circle yet could also participate in a wide range of such circles.

In contemporary mass society, the descendents of the British intellectual journals help to transcend the limitations of geographic spread, large numbers of people and narrow specialization so characteristic of our society. Journals, we shall show, provide for both communications and feedback. Their editors serve in the gatekeeping capacity of the salon hostess deciding who can say what, at what length, to what audience, as well as which books will be talked about and which ignored. Many journals, it will be seen, are the center of lively intellectual circles. Most important, the journals connect an inner core of top intellectuals with a larger audience of intellectually active persons.

By identifying and interviewing those who write for the leading intellectual journals and those whom they name as members of their circles we have identified the American intellectual elite.

Alternative Institutions

There are, of course, a number of intellectual organizations and institutions in the United States today other than journals and the circles which surround them, but in each case they are insufficient in themselves to guarantee certification of the leading American intellectuals. Universities, which once helped to certify leading intellectuals in Britain and Germany, may seem a good starting point for locating the elite intellectuals.[22] Despite some noteworthy exceptions, 90 percent of the leading intellectuals we eventually interviewed have college degrees. More than any other single organization, the university or college acts as the first gatekeeper to the intellectual community, screening out the ineligible. To the extent that colleges emphasize liberal education, they are the guardians of nonspecialized excellence in values and esthetics. Within a single organization, universities house all the specialties which deal with the concerns of intellectuals. Most important, universities and colleges are to a large degree under colleague control.

Yet the growth and proliferation of American universities has doomed them as certifiers of the intellectual elite. Their size and geographic dispersion make difficult the regular interaction necessary for evaluating ideas. There are about 400,000 professors in regular four-year colleges and universities in the United States, and about half of them, according to a study conducted for the Carnegie Commission by Ladd, Lipset and Trow, agree with the statement, "I am an intellectual." This gives us at least 200,000 self-proclaimed intellectuals in the United States, and surely others besides university professors would include themselves in this group. If we are more selective and consider only the 8,000 or so professors in the humanities and the social sciences who have written at least 20 professional articles and who teach in what the Carnegie study called high-quality universities, we find 90 percent of this group saying they are intellectuals.[23] This amounts to over 7,000 persons who may indeed constitute a core of the intellectual community but who are surely too large a number to constitute the very top leadership.

Even if the question of size is set aside, high-quality generalized commentary on economic, social, political, ethical and esthetic problems of contemporary life is not necessarily found among the ranks of university scholars. The university, with its publish-or-perish dictum, tends to reward specialized expertise rather than generalized commentary. Thus scholarly activities, which once led to such commentary, now attempt to copy the value-neutral sciences.

Proof of the lack of interest of many scholars in more general current issues was forced upon us during our interviews for this study. A few persons we talked with were eminent and learned university scholars who had written important scholarly books which had achieved wide notice. Yet almost none of these specialized scholars dared offer firm views on the war in Vietnam, the economy, or any social issue of general interest, pleading lack of knowledge and expertise in matters of public interest. In some definitions of the term these scholars are certainly intellectuals. In our use of the term, the intellectual elite must have and must disseminate to a general audience opinions on matters of values, esthetics, and public policy. So while scholars indubitably are persons of high intellect, they do not necessarily play elite intellectual roles, as we have defined them.

Besides the universities, there are three other types of organization which attempt to certify the intellectual elite: the trade or professional organization, such as the American Sociological Association or the Modern Language Association, the Authors Guild, the National Press Club, and P.E.N.; the official "king-making" honorary societies, such as the American Academy of Arts and Sciences or the National Institute of Arts and Letters; and the semisocial organization or club, such as New York City's Century Club.

Practically all academics belong to one or another of the academic organizations, and 70 percent of those intellectuals whom we eventually sampled belonged to one or another of the other types of organizations. P.E.N. and the Authors Guild were each represented in 30 percent of the sample, the American Academy of Arts and Sciences and the Century Club each included between 20 and 25 percent of the list, the National Institute of Arts and Letters had 10 percent, and the National Press Club 8 percent. There is considerable overlap in the other direction as well: for example, 20 percent of the entire membership of the literature section of the National Institute of Arts and Letters were in our final sample.

None of these organizations does the certification job completely.

Each of the professional or trade organizations engages in some professional functions and some intellectual ones, but even P.E.N., which exercises some selectivity and has broad concerns, is not selective enough for our purposes. None pretends to rank its members or evaluate their ideas. And despite the efforts of "radical caucuses" and the like, American academic societies still concentrate on fairly narrow and technical matters.

The king-making organizations and the social clubs are more selective indeed. But they tend to nominate persons who have already made their mark, generally because of some specialized achievement. Many of their members were never active outside their specialty and others have long since retired from active intellectual combat. Then, too, they have their biases. Judging from the members in our sample, the Boston-based Academy of Arts and Sciences has fewer New York City members than are in the intellectual community as a whole, while the New York–based Century Club naturally has more, as does the National Institute of Arts and Letters. Compared with the Institute, the Academy is more inclined to academics and, coincidentally, the Academy is the only organization mentioned (other than the academic professional ones) in which Jews are represented in the same proportion as they are in our entire sample.[24] The one major benefit and distinction of the king-making groups we can observe statistically is that membership in the Institute of Arts and Letters appears to give a fairly good guarantee that one's books will be frequently reviewed: 75 percent of its members in our sample had had their books reviewed three or more times in the leading intellectual journals we sampled, as compared with 40 percent of members of other organizations.

In sum, if one looks at the list of the top 70 intellectuals given later in this chapter and compares the list with the membership lists of any or all of these organizations, it becomes obvious that all the formal groups include many who clearly do not belong to the intellectual elite and exclude some who ought to belong. Formal organizations are simply less flexible than circles or even journals, for formal organizations always carry the extra burden of organization maintenance. Fifty years of the sociological study of organization can be summed up by saying that organizational goals are almost always subverted from their original purpose by the self-serving needs of organizational life. Intellectual journals and their circles thus remain the one major institution which accurately certifies the top American intellectuals. In the rest of this chapter we shall give the characteristics of American intellectuals which follow from this process of certification. Chapter 2 discusses the journals in more detail and Chapter 3, the circles.

How to Find the Intellectual Elite

Finding the leading intellectual journals is not difficult. If one does not quibble over their relative rank, then almost anyone's list of the twenty or so leading intellectual journals should do. The idea is to include journals of general interest and to exclude specialized or technical ones such as the *American Historical Review*. In 1969, we obtained ratings from a sample of professors, writers, and editors and found that the twenty or so leading American intellectual journals included the *New York Review of Books*, the *New York Times Book Review*, *New Yorker*, *New Republic*, *Commentary*, *Harper's*, *Partisan Review*, *Saturday Review*, the *Nation*, *Atlantic*, *Daedalus*, *Ramparts*, *Yale Review*, *Dissent*, *American Scholar*, the *Hudson Review*, the *Village Voice*, the *Progressive*, *Foreign Affairs*, and the *Public Interest*. Neither the entries on this list nor the various rankings we tabulated within it are permanent.[25] The intellectual bourse constantly changes its listings, and several of the journals on our list obviously had had their ups and downs since the poll was taken; but on the whole the top twenty are still the top twenty.

Once the leading journals are found, locating leading intellectuals is quite straightforward. Our sample of leading intellectuals was drawn from the 8,000 or so persons who contributed to the top twenty-two intellectual journals during the five-year period 1964–1968. The more often a person wrote or the more times his books were reviewed, the more likely he was to be drawn into the sample. Columnists, of course, were handled differently, and editors were automatically included. We also interviewed a dozen intellectuals whom we did not happen to sample but whose names came up very often in the course of interviews with those whom we did sample.

All told, then, our findings about American intellectuals are based on 110 individuals scientifically sampled and interviewed during the calendar year 1970. By our definition, there are probably about 200 leading American intellectuals, so we have more than half of them in the sample. Our "contract" with the persons we interviewed forbids us to reveal their names, but a do-it-yourself attempt on the part of any reader who cares to try our methods informally would probably come very close to reproducing our list of elite intellectuals.

One word more about the list. If there is a prejudice among American intellectuals that quantity does not equal quality, this was not demonstrated in our sample drawn essentially on the basis of quantity of production. For there is a very high relationship between inclusion in our sample and the number of times an intellectual was mentioned as important by other intellectuals interviewed. And the more times a person's books were reviewed in the journals, the more times he was mentioned as important by other intellectuals.

What Do They Do for a Living?

Now that we have indicated how we tracked down the leading intellectuals, we can give an objective portrait of their characteristics. We said that being an intellectual is not an occupation. Just how do intellectuals earn their living? About 40 percent of the 172 intellectuals we tried to interview (there were 62 we never caught up with) are professors, almost 40 percent are editors or staff members of leading journals or newspapers, and somewhat over 15 percent are free-lancers, including several poets and some novelists and short-story writers as well as journalists and essayists. Announcements of the demise of the free-lancer, so loudly bemoaned by so many, seem a bit premature, and the alleged domination of the intellectual elite by academics is also a mistaken impression. It is true that older intellectuals are more likely to become professors; less than 30 percent of those under fifty are professors, but almost half of those over fifty are academics.

Of the academics among the intellectual elite, half are in the humanities but more than 40 percent are social scientists, an occupation relatively new to the ranks of intellectuals. Because journals are so important in certifying the intellectual elite, the group we are talking about has often been called "literary intellectuals." While it is true that the ability to write well and often is important, the presence of so many social scientists suggests that the term "literary" is misleading. And though English professors are the dominant element among the professors in the humanities, they make up only half that group; the rest are split about evenly between historians and a miscellaneous set of fields including foreign languages, comparative literature, and so on. As for the social scientists, they divide into four more or less equal groups: economists, political scientists, sociologists, and a miscellaneous group including anthropologists, geographers, psychologists, and the like.

One can exaggerate the difference between academics and nonacademics among the intellectual elite. (See Table 1.)* Most intellectuals hold several jobs at the same time. Almost all, no matter what their job, engage in free-lance work, and a surprising number of academics are also editors. A few nonacademics teach, and some academics also are newspaper or magazine staff members. When it comes to past jobs, the overlap increases (free-lancing is omitted from our statistics since it applies to almost everyone). Staff jobs are infrequent past history for

* Tables 1–9 will be found at the end of Chapter 1.

academics, but one quarter of the nonacademics have taught, and almost one third of the academics have had positions as editors. Both academics and nonacademics include considerable proportions who have worked in rather prestigious government jobs.

The list of occupations of intellectuals offers strong verification of C. P. Snow's "two cultures" hypothesis. Only one scientist, a biologist, was on our entire list of 172, which also included one science writer who had formerly been a scientist. Had our list of journals included *Science, Scientific American*, or the *Bulletin of Atomic Scientists*, the results might have been different. But these, after all, are specialized journals, one of them the organ of a professional organization. We are not denying the elegance of science or the indisputable fact that it demands a high degree of intelligence and creativity. It is simply that most leading scientists tend to be diffident about communicating their nonexpert personal values and feelings to general audiences. When they do make policy or value pronouncements, these tend to be one-shot affairs in the mass media. Some public work takes place in committees for this or against that (mainly against the Vietnam War, for the preservation of the environment, for the impeachment of Nixon or what-have-you of liberal causes), but aside from their signatures on published petitions, and the adornment their names lend to political pressure groups, most leading scientists can hardly be said to contribute to public debate on such issues. Even where missiles and bombs are concerned, most of the political activity of scientists (and there has been much) is private rather than public and is directed toward influencing high officials rather than toward changing the values of the thinking public. As Deborah Shapley acidly noted in the March 31, 1972, issue of *Science*, "One can climb to high posts, eventually [in the science-government circle] without the encumbrance of having to publicly stand for anything."

The Importance of New York

After occupation, geography, curiously, is the one fact about intellectuals of greatest interest to themselves. They constantly refer to Paris and London as model intellectual centers. Most share the opinion expressed to us by a leading critic:

> I don't think that there is really anything that can now be called the intellectual community in the United States – not in the same sense that one can talk about an intellectual community in Paris or even an intellectual community in London-Oxford-Cambridge. There was for some fifteen years an intellectual community in New York City centered around *Partisan Review* and *Commentary,* and this played to some extent the traditional role of an intellectual community in the capital cities of Europe. But over the past ten to fifteen years the universities have swallowed up most of the people in it, and now you have a spattering of intellectual communities all over the country.

We think intellectual life in America was always more scattered than in Europe. And perhaps it is in comparison with European patterns that New York seems to be on the decline as an intellectual center, though, as will be seen, there are other reasons for this impression. Yet the facts show that there is still a strong concentration of intellectuals in the New York area. Seventy percent of the editorial offices of the top 25 intellectual journals are within 10 miles of the Empire State Building. Over 50 percent of the American intellectual elite themselves are still located within 50 miles of the Empire State Building. (We call this latter range "lunch distance" – the area within which it is easily possible for an author to have lunch with his publisher and return home the same day.) Reflecting American tradition, New England (especially Boston-Cambridge) still has a large concentration of intellectuals, though its 14 percent simply does not put it in the same class with New York. The rest is a scatter, except for Washington, D.C., which, while it cannot rival European capitals, still can boast a surprising 11 percent of the total. (See Table 2.)

New York still serves as a magnet. Among intellectuals who grew up in Europe, New England, Washington, or New York itself, over 60 percent now live in New York; 40 percent of the others now live there too. But New York's preeminence in intellectual life, as well as the rumor that its preeminence is declining, may possibly be traced to the fact that one third of the American intellectual elite actually grew up in New York City.

Since 40 percent of these natives have left New York, and since they were the largest contingent to begin with, it is easy to understand why it is often said that "everyone" is leaving New York. Then, too, a good proportion of elite intellectuals who left New York are academics, and academics have tended to write more about intellectuals than nonacademics. The academics' vision of this world is a bit distorted, for about 55 percent of the nonacademics live in New York, as against only about 40 percent of the academics.

Half of the academics are concentrated at just four universities: Columbia and Harvard each has 16 percent of our academics on their faculties, and Yale and NYU tie for second place with 9 percent each. The figures for undergraduate and graduate school attendance among elite intellectuals show a similar degree of concentration. Harvard was the leading undergraduate school with 12 percent; CCNY had 10 percent, and Yale and Columbia trailed just behind CCNY. Of those who did graduate work, 25 percent went to Columbia, 16 percent to Harvard, and 5 percent to Yale.

Religion, Age, Class and Sex

The dominance of New York is often associated with the supposed dominance of Jews in American intellectual life. Jews are indeed much more strongly represented among leading intellectuals than in the population at large. They compose about half of the American intellectual elite; Catholics are vastly underrepresented, but Protestants, who are one third of the group, are also relatively underrepresented. But there is no difference today between New York and the rest of the country in the proportion of the intellectual elite who are Jews or Protestants – although most of the Catholic intellectuals now live in New York. It is true that three fourths of those in our sample who grew up in New York are Jews, and 90 percent of those brought up in the South are Protestant. Most of those who emigrated to the United States are Jews. But beyond these facts, geography does not seem closely associated with the religious background of intellectuals, most of whom in any case describe themselves today as "indifferent" or "opposed" to religion. (See Tables 4 and 5.)

How large a 50 percent representation is depends on one's point of view. (The proportion of Jews is somewhat higher among the academic segment of the intellectual elite, for 60 percent of the Jews – and only 30 percent of the non-Jews – are professors.) Surely not "everyone" is Jewish, as some intellectuals we have talked with seem to feel. Then there is the question of whom the intellectual elite should be compared with. Certainly, in comparison with the rest of the American elite, the intellectuals are overwhelmingly Jewish in origin. Even in comparison with elite American professors (those who published more than 20 articles in academic journals and who teach in high-quality colleges and universities) of the same age and in the same fields, there are between two and five times as many Jews in the intellectual elite. (See Tables 6 and 7.)

The matter of age brings up the question of the effect of the generation in which they grew up on the character of the elite intellectuals. As with any elite, the American intellectual elite tends to be older than the group from which it is drawn. Only one third of the elite intellectuals who are social scientists are under fifty years of age, as compared with two thirds of elite professors of social science; similarly, one quarter of elite intellectuals who are in the humanities are under fifty, but more than half of the top professors of humanities in the country are under fifty. Taken as a group, one third of the intellectual elite is under fifty, one third is between fifty and sixty, and one third is over sixty. The group over sixty, of course, represents the generation that came of age in the 1930's and early 1940's. The group between fifty and sixty came to intellectual maturity at the end of World War II, while those under fifty grew up intellectually mainly during the Cold War period. (Only 13 percent are forty or under.) Surprisingly, the effect of these generational differences upon the opinions of intellectuals is negligible (see Table 8), though as will be seen in Chapter 3, Page 90, there is a generation effect on the opinions held by various intellectual circles.

Of course, one reason for the greater average age of elite intellectuals is that their establishment had not been revolutionized by World War II. The professorial elite came to its position not solely because they climbed upward in a static system but because the system itself expanded after World War II. The professors are relatively young, like the elites of developing nations, and like these elites, they are now beginning to face the clamor of those still younger, who find their paths blocked by an elite that is likely to remain on the scene for many more years. Because the intellectual elite is relatively old, there is room at the top, although the process of getting there may have changed.

A study of how one becomes an elite intellectual is beyond the scope of this book. The first draft of the interview used in this study devoted several questions to the matter of career and "making it"; answers ran over an hour and so – except for actual job history – we reluctantly abandoned any attempt seriously to inquire into the making of an intellectual. Despite our lack of solid data, the paucity of young intellectuals in this sample may well be caused by the fact that recruitment to the role of elite intellectual operated differently in the 1960's than it did in the thirties, forties and fifties. I have the impression from reading autobiographical accounts of intellectual life in these time periods (for example, Alfred Kazin's *Starting Out in the Thirties,* or Podhoretz's *Making It*) that young intellectuals tended to be sponsored by older intellectuals into intellectual prominence through a combination of journals, circles, and political parties controlled by the older intellectuals. This combination faltered in the sixties when young would-be intellectuals tended to reject existing journals, circles, and parties and indeed rejected the very idea of a life built mainly on the foundation of intellect as opposed to action and feeling. Along with this rejection of existing forms and institutions went an attempt to build new ones, but these institutions proved ephemeral. In 1970 when we collected our data, there were practically no institutional bases from which we could draw a sample of young intellectuals and, as will be seen, the older elite intellectuals were systematically ignorant of who the prominent young intellectuals were. In some ways, then, the sixties were the occasion for another "lost generation" of intellectuals, albeit lost in quite a different way from the generation of the twenties.

The young intellectuals of the sixties, therefore, do not show up in appreciable number in our sample, though indeed we have interviewed some of them, and this lack may account for our relatively unimpressive findings on the effect of generation on the opinions of leading American intellectuals. But this finding and the fact that there are few young elite intellectuals is not an "error": it reflects the structure of intellectual life in the United States at this time.

To return to the related topic of religion or ethnicity, the post–World War II expansion of universities and colleges and the accompanying relaxation of barriers against Jewish professors among the elite led to the presence of twice as many Jews in most fields as before the war. Science and the professions of law and medicine are exceptions because Jews were already well represented among the elite before the war.[26] In contrast, among the intellectual elite easier access for Jews began in the 1930's and early 1940's; among intellectuals over sixty years of age there are only a few more than one third who are Jews, whereas among those in younger age groups about half are Jews.

In one area generation does make a difference. Throughout all of western Europe there has been a tendency in the twentieth century for intellectual careers to be more and more open, and talent, whatever a person's class origin, has tended more and more to be recognized. And so today more elite intellectuals come from lower-class backgrounds than do other Western elites, for whom class is a more substantial barrier. (The very opposite holds true for eastern Europe – intellectuals there have higher class origins than the rest of the elite.) This trend is observable even among American intellectuals, for while barely over 10 percent of the fathers of elite intellectuals over sixty came from a manual-labor background, 40 percent of the fathers of younger elite intellectuals were manual workers.

Becoming an elite intellectual implies for many substantial upward mobility. When it comes to current class position, the elite intellectuals are indeed an elite. Their median family income in 1969 was about $35,000, considerably more than that earned by leading university or college professors. Of course, intellectuals who are at the top of their heap earn a good deal less than those at the top of the heap in other institutional areas, such as business or even politics.

Finally, we have to note that very few women indeed were among the top American intellectuals in 1970.

Liberal – But Not Radical

Given their background and position, just where do elite intellectuals stand politically in comparison to everyone else? To the left. The American intellectual elite is more liberal on any issue of public policy than the American public at large, more liberal than any other segment of the American elite,[27] and generally more liberal than the elite university and college professors surveyed in the Carnegie study (with appropriate controls for age, field, and religion). (See Table 9.) In an important review of the politics of American and Soviet intellectuals, Lipset and Dobson in the Summer 1972 issue of *Daedalus* conclude, on the basis of the Carnegie study of American professors and other data, that intellectuality per se seems to make a person more liberal and more critical of the policies of the regime.[28] If our sample is more "intellectual" than even the elite among the Carnegie study sample of American professors, then we have additional verification of their hypothesis. It is being an intellectual per se that counts. Except on some of the older issues related to socialism and capitalism, where Jewish elite intellectuals tend to be farther to the left than others,[29] social class background and even age are unrelated to opinion among the intellectual elite. If anything, elite intellectuals from working-class backgrounds are more conservative than the rest; younger members of the elite intellectuals act as if they had grown up in the 1930's with the then radical generation; and non-Jews vote almost 100 percent Democratic or third-party just like the Jews. The elite intellectuals have so long been involved in the culture of intellectuals that their past backgrounds have become almost irrelevant.

But this is not to say that most of the intellectual elite are radical. They are not, as we shall observe at length in Part II. In 1970, less than half wanted to get out of Vietnam immediately, even though most intellectuals had strongly opposed the war since 1965, long before any other group in the population. And their opposition remained mainly on pragmatic grounds – the war did not work – rather than on ideological or even moral grounds. And in Part III we will see that a large number of intellectuals are actively concerned about race, but most oppose "Black Power" and most frankly admit they do not know how to solve America's race problems. The cultural crisis and the changing values of youth are of even more concern to the intellectuals than is race, but most of them feel that the young are way off base. Only a few of the intellectual elite are devoted today to the overthrow of "capitalism" as a system, though one third described themselves as "radical" in their youth. It is for these reasons that some of the younger academics have attacked the intellectual elite as defenders of the Establishment.

The Top Seventy

Now that we know something about the social characteristics and the political opinions of elite intellectuals, we can elect the more elite from the ranks of the mere elite. The election is not our doing, of course, and it follows once again the principle of colleague certification. We asked the elite intellectuals in our sample to name intellectuals who influenced them on cultural or social-political issues, or who they believed had high prestige in the intellectual community they felt themselves to be a part of.*

* Here are the actual questions. Not every respondent was asked every question. The aim was to find *something* he could comfortably answer.
1. What intellectuals who write on social-political issues have the highest prestige in the intellectual circle or circles that you are closest to?
2. What intellectuals who write on social-political issues do you think have the highest prestige in the intellectual community in general?
3. What intellectuals who write on cultural matters have the highest prestige in the intellectual circle or circles that you are closest to?
4. What intellectuals who write on cultural matters have the highest prestige in the intellectual community in general?
5. What intellectuals who write on both social-political and cultural matters have high prestige in the intellectual community as a whole?
6. Could you tell me the names of the persons whose views on social-political issues in general have most influenced your own thinking?
7. Could you tell me the names of those people whose general views on cultural issues have most influenced your own?

A number of different questions were asked, but each person who was polled could register only a single "vote" for a particular intellectual, no matter how many different questions he answered with the same name.[30]

Several caveats about our list of top intellectuals are in order. First, it applies to 1970, when the interviews were undertaken, and represents the consensus of the leading intellectuals of the late 1960's, not the early 1970's. Second, the questions were asked in the context of an interview about Vietnam and about American social issues. Vietnam experts may thus appear more prominent than they might otherwise have been. Despite the fact that half the direct questions asked for cultural leaders and half for social-political ones, respondents may have been influenced by the context to slight literary persons and to emphasize social scientists. (There are a good number of social scientists among those interviewed, however, and the growth in importance of social science among intellectuals may be greater than has generally been recognized.) Third, not everyone was happy with our request for names. A leading fiction writer commented:

> I've been asked this question [who are important writers] recently by *Esquire*. I feel very strongly that writing is not a competitive sport, and as soon as you start grading writers as being more important or less important than other writers, then you completely misunderstand the fact that all writers are very much dependent on one another. This includes the bores, the fools. I've known unsuccessful writers, X for example, whose books were virtually unreadable, but whose contribution as an enthusiast and a conversationalist is of inestimable value. I don't think you can rank writers. I don't think you can grade them. As I say, we are all very dependent upon one another. One can't simply say that Saul Bellow is better than Ralph Ellison or that Norman Mailer is better than Kurt Vonnegut. I mean, it doesn't make any sense, because they are all very closely dependent on each other. It's a creative climate, an intellectual climate in which people can work freely and with pleasure.

We agree, of course, and that is why we have stressed the notion of a circle of intellectuals. Even with his objections, however, this respondent gave the names of intellectuals he considered important in various ways. Almost everyone else did too.

Following the precedent established by Moses, we hereby list the 70 "elders" of the American intellectual camp, as named by our "jurors." They are listed alphabetically within four general ranks.

The reader is welcome to try a do-it-yourself analysis of the list. One should note that leadership here is not the same as "getting around" a good deal, and hence not the same as centrality in a social circle, though it is of course related. A more accurate analysis is possible by comparing those with more

The Seventy Most Prestigious Contemporary American Intellectuals (1970)

---

*Ranks 1 to 10 (two tied for tenth place)**

Daniel Bell
Noam Chomsky
John Kenneth Galbraith
Irving Howe
Dwight Macdonald
Mary McCarthy
Norman Mailer
Robert Silvers
Susan Sontag
Lionel Trilling
Edmund Wilson

Ranks 11 to 20

Hannah Arendt
Saul Bellow
Paul Goodman
Richard Hofstadter
Irving Kristol
Herbert Marcuse
Daniel Patrick Moynihan
Norman Podhoretz
David Riesman
Arthur Schlesinger, Jr.

Ranks 21 to 25 (numerous ties)

W. H. Auden
Norman O. Brown
Theodore Draper
Jason Epstein
Leslie Fiedler
Edgar Friedenberg
John Gardner
Eugene Genovese
Richard Goodwin
Michael Harrington
Pauline Kael
Alfred Kazin
Murray Kempton
George Lichtheim
Walter Lippmann
Marshall McLuhan†
Hans Morgenthau
I. F. Stone
C. Vann Woodward

Ranks 26 and 27 (numerous ties)

Edward Banfield
Isaiah Berlin†
Barbara Epstein
R. Buckminster Fuller
Nathan Glazer
Elizabeth Hardwick
Robert Heilbroner
Sidney Hook
Ada Louise Huxtable
George F. Kennan
Christopher Lasch
Seymour Martin Lipset
Robert Lowell
Robert K. Merton
Barrington Moore, Jr.
Willie Morris
Lewis Mumford
Reinhold Niebuhr
Robert Nisbet
Philip Rahv
James Reston
Harold Rosenberg
Philip Roth
Richard Rovere
Bayard Rustin
Franz Schurmann
John Simon
George Steiner
Diana Trilling
James Q. Wilson

* Within ranks listed in alphabetical order.
† Though McLuhan is a Canadian and Berlin an Englishman, both men have spent enough time at American universities to warrant their inclusion on our list.

"votes" to those with fewer votes, utilizing the entire list of 172 intellectuals. We wish to emphasize that for administrative reasons not everyone in the top 70 was on our list to be interviewed, so there is no guarantee that inclusion in the list means that we tried to interview the person, much less that he was actually interviewed. But it is also true that our sample includes a large number of important intellectuals, and that among the 172 on the list, the sample of those interviewed is biased only in terms of the relative paucity of important free-lancers. These are in short supply mainly because they do not have secretaries who force them to meet their social obligations. Presumably, this is precisely why they are free-lancers. Only the top 70 are shown here. Actually, two thirds of the members of the list of 172 received fewer than two "votes," 16 percent between two and five votes, and 18 percent six or more.[31]

The first surprise is that age is *not* correlated with the number of votes. The reason for this is probably that much older intellectuals tend to slip a bit, especially since our ratings give points for being around and for being active discussion partners. However, younger intellectuals do not make the list at all unless they are quite exceptional. The net result is that age does not affect an elite intellectual's rank. Another surprise is that living in the New York metropolitan area (our fifty-mile district) is *not* positively related to prestige. The New York "Mafia" is seriously overrated.

Some stereotypes about intellectuals do seem to be borne out by the characteristics of intellectuals with the most votes. More than 40 percent of the Jews received six or more votes as compared with about 15 percent of the non-Jews. Forty percent of the humanities professors received six or more votes, while only 25 percent of nonacademics and social science professors had six or more votes. Radicalism in one's youth is strongly associated with prestige; none of those who said they were conservatives or middle-of-the-roaders in their senior year in college received six or more votes, but 60 percent of those who were college radicals did. The association between the radicalism of one's current views and prestige is of much more modest proportions. (On the big issue of the day, the war in Vietnam, there was no difference in votes between those who wanted to get out immediately and those who favored a coalition and more gradual withdrawal. On the other hand, very few of those favoring preventing a communist takeover got any votes at all.) And on the whole, the number of contemporary "radicals," self-proclaimed or otherwise, in the top seventy is small indeed.

Social circles of intellectuals are very much a historical phenomenon – their characteristics are highly related to the time and place in which they appear and the institutional pegs upon which they hang. The elite American intellectuals as we saw them in 1970 were basically the same ones who came to power in the late 1940's and early 1950's, as is shown by a study by Tom Conway. Their social characteristics and their political orientations are still tied to the circumstances and issues of the Great Depression, Nazism, and Stalinism and anti-Stalinism. On current issues they are not necessarily avant-garde, though they are more liberal than comparable groups. The war in Vietnam and revisionist views of the Cold War did succeed in "electing" some new faces, but despite the impression some have of the *New York Review of Books,* the new organ of the intellectual elite, such persons have not made a major dent in the ranks of the elite intellectuals. The problems of race relations have brought in some Blacks, but none have been especially new to the scene. The "cultural revolution" (drugs, rock music, and the like) and the new style of radical politics also brought in a few new faces, but by now most have dropped out of major intellectual social circles.[32] If the generation under thirty has formed new circles, they are not yet cohesive enough for us to identify and find them. If they exist at all, they exist in isolated pockets, not national networks, and in 1974 we would be hard pressed to change radically our 1970 sample of intellectuals.

Symptomatic of this confusion is the inability of most intellectuals we interviewed to name their replacements. Some have felt that the whole idea of an intellectual elite is passing. I doubt it. The patient is quite alive and well, thank you, but he is getting older and his heirs have not yet made themselves known.

TABLE 1

Additional Jobs or Occupations of American Intellectuals (In Percentages)

SECONDARY OCCUPATION	NONACADEMIC (N = 94)	PROFESSOR, HUMANITIES (N = 31)	PROFESSOR, SOCIAL SCIENCES OR OTHER (N = 32)
Present			
Academic	6	0	0
Editorial	38	19	16
Magazine or newspaper staff	29	9	3
Free-lance	87	81	78
Government	8	0	0
Past			
Academic	24	91	97
Editorial	60	31	34
Magazine staff	35	6	6
Newspaper staff	42	3	3
Government	25	28	34

NOTE: Percentages add to more than 100 because many hold or have held several jobs simultaneously.

TABLE 2

Geographic Location of American Intellectuals (In Percentages)

REGION	CURRENT LOCATION (N = 163)	LIVING IN HOMETOWN[a] (N = 160)
Outside U.S.A.	2	12
New England	14	11
Mid-Atlantic Region (excluding New York City)	5	14
South (excluding Washington)	4	9
Midwest	6	12
West	7	8
Washington, D.C.	11	2
New York City area[b]	51	32

[a] Attended elementary school there.
[b] Fifty-mile radius.

TABLE 3

Geographic Background of Intellectuals from Elsewhere Now in New York City

REGION	PERCENTAGE
Outside U.S.A.	55
New England	67
Mid-Atlantic Region (excluding New York City)	41
South (excluding Washington)	33
Midwest	37
West	33
Washington, D.C.	100
New York City area[a]	58

[a] Includes fifty-mile radius.

TABLE 4

Religion in Which Intellectuals Grew Up

REGION IN WHICH INTELLECTUAL GREW UP	RELIGION (PERCENTAGE)					
	PROTESTANT	CATHOLIC	JEW	OTHER	NONE	NUMERICAL TOTAL
Outside U.S.A.	24	6	64	6	0	(17)
New England	54	15	15	15	0	(13)[a]
Mid-Atlantic Region	47	0	42	0	11	(19)
South	92	8	0	0	0	(13)
Midwest	44	11	11	17	17	(18)
West	36	18	28	0	18	(11)
Washington, D.C.	67	0	33	0	0	(3)
New York	15	2	81	2	0	(47)

[a] In this table and those that follow, percentages may add up to 99, 100, or 101 depending on rounding.

TABLE 5

New York and Non–New York Intellectuals and the Religions in Which They Grew Up (In Percentages)

RELIGION	CURRENT RESIDENCE	
	NEW YORK	NON–NEW YORK
Protestant	31	34
Catholic	12	3
Jew	44	49
Other	6	6
None	8	8

TABLE 6

Elite American Professors Brought Up as Jews

FIELD	AGE			
	UNDER 50		50 AND OVER	
	%	N	%	N
Social Science	39	(2875)	21	(1263)
Humanities	24	(2248)	12	(1767)
Science	26	(6141)	28	(1783)
Professions	33	(2718)	28	(1220)
Other	21	(4729)	10	(2979)

TABLE 7

Elite Intellectuals Brought Up as Jews

FIELD	PERCENTAGE	NUMBER
Social Science	56	(30)
Humanities	61	(28)
Nonacademic	29	(80)

TABLE 8

Elite Professors and Elite Intellectuals Under Fifty Years of Age, by Field

FIELD	ELITE PROFESSORS		ELITE INTELLECTUALS	
	%	N[a]	%	N
Social Science	67	(4138)	34	(32)
Humanities	56	(4015)	26	(31)
Science	78	(7924)	[b]	
Law or Medicine	69	(3939)	[b]	
Other	61	(7708)	[b]	
Nonacademic	[b]		37	(94)

[a] Weighted to give the numbers of professors in the United States in these categories who teach in high-quality universities and colleges and who have written 20 or more articles.

[b] Insufficient or no representation.

TABLE 9

Opinions of Elite Intellectuals and Elite Professors,[a] *According to Academic Field (In Percentages)*

QUESTION	SOCIAL SCIENCE: INTEL-LECTUALS (N = 21)	SOCIAL SCIENCE: PROFES-SORS[b]	HUMANITIES: INTEL-LECTUALS (N = 10)	HUMANITIES: PROFES-SORS[b]	NON-ACADEMIC: INTEL-LECTUALS (N = 32)
I am intellectual	100	91	100	89	75
Universities					
Disrupting students should be expelled	60	71	60	79	35
Militant students are a threat to freedom	85	81	60	81	52
Left causes the trouble	30	40	40	44	23
Blacks					
Blacks should control own schools	86	60	70	58	87
White racism is cause of riots	55	56	30	46	63
New Politics					
Politics can't make for change	24	16	40	28	21
There is no justification for violence	57	62	60	65	66
Marijuana should be legal	85	60	70	39	93

TABLE 9 (*cont.*)

Opinions of Elite Intellectuals and Elite Professors,[a] *According to Academic Field (In Percentages)*

	FIELD				
	SOCIAL SCIENCE		HUMANITIES		NON-ACADEMIC
QUESTION	INTEL-LECTUALS	PROFES-SORS[b]	INTEL-LECTUALS	PROFES-SORS[b]	INTEL-LECTUALS
Foreign Policy					
Communism makes for progress in developing countries	5	14	10	27	17

[a] Because the compositions by religion and age of elite professors are so different from that of elite intellectuals, and because among professors these characteristics are strongly related to expressed opinion, the data for professors has been restated to represent their opinions assuming they have the *same* age and religious distribution as the elite intellectuals. The issues shown were represented by questions more detailed, obviously, than the captions on the table, and were administered to professors in the Carnegie Commission for Higher Education study, and by courtesy of the commission and Professors Lipset and Trow were identically copied for the study of American intellectuals one year later in 1970. The questions were administered to intellectuals via questionnaires returned after a two- to six-hour interview. This was a mistake because intellectuals, especially in the Humanities, do not like questionnaires. We got only 60 percent of them back. The chief bias is the low rate of return by litterateurs, corrected by presenting data for social scientists, humanities professors, and nonacademics separately.

[b] As explained in note a, the number has been adjusted. Original numbers are shown in Table 6.

2

The Power and the Glory: The System of Intellectual Journals

Contemporary American intellectual journals are the chief agencies for grading the worth of an American intellectual and are mainly responsible for the prominence of the top seventy whose characteristics we have just reviewed. A leading American intellectual recently irked some of his colleagues by openly confessing that success was of central importance to him.[1] Podhoretz, disappointed at the reaction to his book *Doings and Undoings,* finally worked up the nerve to measure his stock of prestige as an intellectual and, as he says,

> discovered to my surprise that after much active trading my stock had registered an impressive gain. It had started moving hesitantly with a page-four review (and picture) in the *Sunday Times;* rose again slightly with a picture and a flattering piece in *Newsweek;* fell off somewhat after a vicious assault on me in *The New York Review of Books* (unaccompanied by a Levine caricature, which cost me another point); but then rose sharply when the usual honor of being attacked at great length in the *New Yorker* was paid me in spite of their long-standing rule against reviewing books containing material, as *Doings and Undoings* did, which had originally appeared in their

A somewhat different version of this chapter, by Julie Hover and myself, appeared in the March, 1972, issue of *Change* magazine. Hover is mainly responsible for the analysis reported here and I for the present formulation.

pages; went up still further when *Partisan Review* featured their piece about me on the cover; and eventually leveled off at a new high as the publication date of the book receded into the past.[2]

To understand the world of the leading intellectuals of the late 1960's, then, one must first understand something about leading intellectual journals of that time. To begin with, we should like to know which were the leading journals and what made them influential. Then we should like to review the relationships between journals and intellectuals, including who makes whom. Our elaborate ratings of the influence of journals and individual intellectuals is, at one level, mere organized gossip, interesting as that may be. But how ideas are selected and transmitted, who determines which ideas are to be influential as well as who may pronounce them – these are important to political and cultural life.

What Is Being Read

To be influential, a journal must be read, preferably by prestigious intellectuals. As shown in Table 10, among the eight journals (from a checklist of 42) ranked highest by our sample in 1970, the *New York Review of Books* and the *New York Times Book Review* were tied for the first place as "most read," with about 75 percent of the leading intellectuals claiming to read them regularly. (The *New York Times Magazine* is actually in first place, with over 80 percent, but it is a special case, to be discussed below.) In second place, with over half the sample claiming regular readership, were the *New Yorker, New Republic, Commentary*, and *Harper's*. In third place was *Partisan Review,* with one-third readership. *Saturday Review* was read by slightly more than 20 percent. The entire list follows below in alphabetical order. (This ranking is somewhat different from that reported in our first study of a less selective set of respondents, which was designed to find the top journals so that we could choose our sample of writers.)[3]

Professors, we found, read fewer of these journals than nonacademics; they averaged 11 journals apiece, from the 42-

TABLE 10

Journals and Magazines Regularly Read by 90 Elite American Intellectuals in 1970

JOURNAL OR MAGAZINE[a]	PERCENTAGE READING "REGULARLY"[b]
American Scholar	22.2
Atlantic	47.8
Atlas	16.7
Columbia Review	18.9
Commentary	54.4
Commonweal	18.9
Daedalus	41.1
Dissent	38.9
Esquire	25.6
Foreign Affairs	26.7
Harper's	53.3
I. F. Stone's Bi-Weekly	24.4
Life	28.9
Look	20.0
Nation	22.2
National Review	16.7
New American Review	12.2
New Republic	54.4
New York Review of Books	74.4
New York Times Book Review	78.9
New York Times Magazine	81.1
New Yorker	58.9
Newsweek	52.1
Paris Review	13.3
Partisan Review	32.2
Playboy	10.0
Progressive	10.0
Public Interest	27.8
Ramparts	16.7
Saturday Review	23.3
Time	46.7
Transaction	22.2
Village Voice	22.2

[a] Journals listed are those read by at least 10 percent of the 90 respondents who answered the question.

[b] The actual wording of the question explained, "That is, read something in most issues."

journal list, while nonacademics averaged 14 journals each. Of course, these results may reflect the fact that professional journals, probably read more often by academics, were not included on our checklist. Those persons mentioned two or more times by the rest of the sample as being influential intellectuals read more journals.

The influential nonacademics claimed, on the average, to have read regularly a whopping 16.5 journals. Even the slightly more than 9 journals the less influential academics read, the dozen journals influential academics read, or the slightly more than 11 journals the less influential nonacademics read seem to be very large numbers indeed. One almost gets an impression of an opinion-making machine which, in order to produce top-quality opinions, must be fed with a large ration of top-quality opinions. In fact, the greater the prestige of an intellectual, the more of the top journals was he likely to read.

Part of the reason for the greater number of journals read by nonacademics may be their desire to keep up with the interests of the general public, a matter of apparently less importance to the academics. A greater proportion of the nonacademics read the mass-circulation publications such as *Newsweek, Time, Life* and *Esquire,* while more of the professors read those with smaller circulation such as *Daedalus, Dissent, The Public Interest* and *Commentary.* But both academics and nonacademics agreed on five of the six journals they considered influential. The *New York Review of Books,* the *New Republic,* the *New York Times Book Review,* the *New Yorker* and *Harper's* are read by a large proportion of both groups. *Commentary,* published by the American Jewish Committee, seems to be slightly preferred by the academics, which may be because more of them are Jews.

The *New York Times Sunday Magazine* is a special case. Literally 98 percent of the intellectual elite interviewed (including those who do not live in New York) said they read the *New York Times* and more than three fourths of them also said they read the *New York Times Sunday Magazine.* Although the *Times Magazine* had not been included in a first survey because it is technically classified in directories as a Sunday supplement, it was included in the expanded list of journals used in the present study so that it might be compared with the journals. Although this magazine is widely read, there is little evidence from the comments in the interview that it is especially influential among intellectuals as a separate publication. Some said the *New York Times Book Review,* on the other hand, can make or break an intellectual's reputation.

Who Writes for What

The intellectual world is a circular one. Leading intellectuals write for the leading journals because, even without the benefit of surveys, they believe that these journals are read by the intellectual elite. So that when an issue comes up that really disturbs them, such as the war in Vietnam, they take up the pen (or bang the typewriter) and write for intellectual journals: 60 percent of our sample wrote on the war.

We asked the intellectual elite how they would (or did) get their ideas on the war in Vietnam across to other intellectuals. The large majority said they would write in journals; the more influential elite were more likely than the less influential to say they would write. Slightly over half of the intellectual elite mentioned specific journals and most people who mentioned journals named at least two. The journal selected most often as the medium for reaching other intellectuals was the *New York Review of Books,* selected by 21 persons. Other journals named by at least 10 persons were *Commentary, Sunday Times Magazine, Atlantic* and *Harper's.* Journals mentioned from 5 to 8 times were *New Republic, Dissent, Partisan Review, New Yorker* and *Nation.* There was a tendency for the more influential elite intellectuals to name a greater number of journals and to name influential journals most often. For example, 16 of the 21 who selected the *New York Review of Books,* and 14 of the 17 who selected *Commentary* were mentioned twice or more by our sample as being influential intellectuals. Although these more influential intellectuals were likely to state exactly how they would get their views across to intellectuals, less influential intellectuals were less likely to know which journal to use.

A statistical list of intellectual journals suggests that the top ones are all useful for reaching intellectuals. But the more influential intellectuals tended to think that there was one organ for intellectuals — the *New York Review of Books:*

If and when I write for my fellow intellectuals, I write for the *New York Review of Books.*

A number of persons who wrote in the *New York Review of Books* have made the *New York Review* a kind of organ of anti-Vietnam, intellectual sentiment.

In reaching intellectuals, the *New York Review of Books,* which I disagree with, is certainly a very important mailbox.

When other journals were selected, they were usually perceived as less important in reaching intellectuals. The following illustrative comments were made by an intellectual who does not write for the *New York Review of Books:*

There are no mass publications for intellectuals other than the *New York Times Book Review* and *Magazine* sections. Both reach intellectuals as well as the mass public. There are two really effective places for reaching intellectuals and general readers . . . *New Yorker* . . . *Harper's.* The most effective, the most powerful intellectual journal without question is the *New York Review of Books.* No question about that being so. It is not a place I can write, but if you want to reach the intellectual community, that's the medium to use.

In selecting journals in which they would write, some intellectuals took into account the type of audience. Interesting feedback was very important. Intellectuals felt that articles published in influential journals resulted in more serious and more frequent feedback from readers than articles published in less influential journals. As one leading writer put it "[based on letters I receive] . . . I have noticed that if I write something in the *Saturday Review,* only uninteresting people read it, but if I write for the *New York Review,* everybody reads it." Another intellectual told about an acquaintance who got more letters in response to one book review appearing in the *New Republic* than he had received in all the years he had written columns in two noninfluential journals. Still another stated that he got far more letters when he wrote for the *New Yorker* than he got when writing for *Esquire,* even though the circulation of *Esquire* is much greater. Other differences in audiences were noted. One intellectual, for example, pictured the *Harper's* audience as "the earnest housewife" but couldn't figure out "who the hell reads *Atlantic.*"

Sheer mass circulation does have its appeal, however. One leading critic for a top-rated journal would like on occasion to reach a larger audience:

> If I did a long, analytical article on the state of the theater . . . *Harper's* or *Atlantic* wouldn't publish it. The editor might love it, but he wouldn't publish it because they are very much in sweaty, competitive business these days and they use intellectual subjects that are hot in the topical sense or that are written by someone sensational. They would print an article on ecology. If Andy Warhol wrote a piece about nakedness in the theater, *Harper's* and *Atlantic* would break their back to get it. But if Eric Bentley or Francis Ferguson wrote a brilliant article analyzing the current philosophies of the theater, the editor might read it with fascination and say, "You know this would just turn them off like crazy." Magazines like *Harper's* . . . they're not really outlets for intellectual expression. They're usually counted in, and they're not. *Esquire* is not. It wants the hot thing. Therefore, you get this real schism in this country between the large circulation and small circulation magazines. And therefore you get yourself painted into the box or corner of talking to people who probably already agree with you.

The large-circulation journals tend to create and to cater to intellectual celebrities whose products may not have received the careful scrutiny by the circle of peers to which most of the work in intellectual journals is subject. The "hot thing" and the star system together conspire to produce Erich Segals and Charles Reichs, neither of whom, it should be noted, appears on our list of the top seventy.[4]

Most of the sample, however, did not know how to reach the general public and/or did not care to do so. When asked how they would get their views on Vietnam across to the voting public, the intellectuals gave rather vague answers. Fewer than half said they would write, with the majority specifying journals rather than books as the medium of choice. The next most frequently suggested method was television, mentioned by about one fifth, followed by writing in newspapers, mentioned by only 10 percent. Some intellectuals said they would not try to reach the public at all, explaining that they wrote only for intellectual audiences. One "never considered that sort of thing," while another retorted that writing for a general public was "alien to my kind of thinking." Another explained, "I don't ever address the general voting public. My communication is with other intellectuals."

Once having said they would write for the general public, most intellectuals were quite vague as to how they would go about it. Only three quarters could mention a specific journal and only five could name more than one journal. The journal most often mentioned as best for reaching the general public was the *Sunday Times Magazine*, but this was mentioned by only seven persons. None of the intellectuals would try to reach the public by writing in the *New York Review of Books*, *Commentary* or *Partisan Review*, and only one to three persons selected *Harper's*, *Atlantic*, *New Republic*, *New Yorker*, or the *Nation*. According to the replies of most of our respondents, influential journals are obviously no way to reach the general public directly.

Some intellectuals held to a "trickle down" theory of influencing the public, however. An editor of a major intellectual journal, albeit one with a relatively small circulation, explained to us his theory of influence:

> The journals of opinion . . . *New Republic*, *Nation*, the *New Leader* . . . influence other publications very strongly. You'll see the popular mass magazines pick up an issue a year after the journals of opinion have had it. . . . Somebody told me that there used to be a woman who largely did nothing for Henry Luce except watch the smaller journals and mark and clip things that would be of interest.

And another editor of a leading intellectual journal amplified:

> To a very large extent small circulation journals germinate the ideas that are disseminated out to society. They prepare the essential

agenda. . . . It is not a direct influence, it is primarily indirect. It works on the thinking of people in charge of editorial pages, writers, other publications, the leaders of thought, most of the leaders and all kinds of opinion leaders.

Intellectuals who would not or could not write as a means of influencing anyone had their own theory of Modern Times. Television is the true source of influence, as a leading intellectual told us:

Writing isn't so important any more. People in this country who are forming opinion are not these old intellectuals who are writing for these old journals. They're the people who are broadcasting over networks . . . the kickback from something that is said on the Johnny Carson show to Congress is very superior to what appears in the *New York Review of Books.*

What Makes for a Top-rated Intellectual Journal

Whatever the perceptions of intellectuals about the leading journals, we have the data with which to construct a "formula" for a successful intellectual journal. Two crucial ingredients separate the more influential journals from the less influential ones: what they publish and whom they publish. We analyzed the content of the top 24 journals for the years 1964 through 1968 (see the Appendix on Sampling for the list of the top 24). This period coincides with the peak trends of the 1960's and so affords a look at a world that is already regarded with some nostalgia by leading intellectuals.[5]

The late sixties are already being acclaimed (or mourned) as the age in which American intellectuals became politicized and moved to the left. Our data suggest (as will be seen) that this is not quite true of our sample. Nonetheless, a move to the left was especially noted with regard to the intellectuals' leading publication, the *New York Review of Books.*

In a sense it is true that every mass circulation magazine has turned to the left, but it's fluff. They are competing for an affluent, youthful public. When the *New York Review of Books* began, I wrote for it quite a good deal. Then it began to be clear to me in the last few years that they were diverging in a direction that I did not like – printing people like Kopkind, whom I abominate, and Tom Hayden and other such. They were swinging in with the New Left fashion.

What I normally read has become increasingly preoccupied with Vietnam. For example, I read a good deal of the *New York Review of Books,* which has articles about Vietnam running out all over the pages.

The *New York Review of Books* has a great deal of power and influence on the literary scene in New York. Of course I say literary and that is the way it began. It wound up being far more a political and social journal than it has a literary one.

Others feel this is not true. People close to the *New York Review of Books* have pointed out that it had substantial sociopolitical content right from the start in 1963. This may be true, but an accurate content analysis over the five-year period we looked at reveals that *all* journals which had a relatively high proportion (20 percent or more) of literary content in 1964 and 1965 have since greatly decreased the space allotted to literary topics. The one exception, *Kenyon Review,* has ceased publication. This decreased attention to literature was accompanied by an increased focus upon the arts and upon sociopolitical issues. Beginning in 1967, the six journals most read by our sample – the *New York Review of Books,* the *New Republic, Commentary,* the *New York Times Book Review, New Yorker* and *Harper's* – *all* showed a marked increase in sociopolitical content with a corresponding decrease in content concerning literature and, in this case, also in the arts. And this is what sets off the more influential journals from the rest. Only 5 of the 18 less influential journals increased their sociopolitical content during this time period.

Part, but by no means all, of the increase in sociopolitical coverage was caused by the increased number of articles and reviews of books on Vietnam and, to a lesser extent, on college campus crises. Again, 5 of the 6 influential journals increased the proportion of space allotted to hot sociopolitical issues such as Vietnam and student unrest, whereas only 5 of the 18 less influential journals followed this trend. The one exception among the influential journals was *Commentary,* already very high in "hot" issue coverage.

When asked which journals had most influenced the interviewee on Vietnam, *New York Review* was frequently mentioned as having a *negative* influence on the thinking of the respondent, especially because of its left content. Other journals frequently mentioned, though rarely because of their negative influence, were the *New Yorker, New Republic,* and *I. F. Stone's Bi-Weekly,* followed by *Harper's, Commentary, Dissent* and *Atlantic.* The *New York Times Sunday Magazine* was also mentioned frequently, with a number of persons pointing out that some of the *Times* coverage was conservatively biased. In general, journals regarded as influential or as having good coverage were those whose reporters wrote from the scene of action abroad and switched to an anti-Vietnam position but were self-conscious about this change. Any on-the-spot coverage was also regarded as important.

Although an *increase* in political coverage coincides with more influence, intensive political coverage, per se, is no guarantee of great general influence. In fact, the three journals with the highest proportion of political content (*Progressive, Dissent* and the *Reporter*) were not among the most influential. Similarly, the three journals with the highest proportion of literary content (*Yale Review, Kenyon Review* and *Sewanee Review*) failed to make the influential list. The influential journals, with the possible exception of the *New Republic,* do not have a large proportion of their content in any single category. *General coverage* is therefore a very important ingredient in becoming a top influential journal.

If our view of journals as the gatekeepers of conversation as well as fame in the intellectual world is correct, then book reviewing should be an important function, for that is the quintessence of gatekeeping. The more important a journal is, the more book reviews it should have. Extreme examples are the *New York Review of Books* and the *New York Times Book Review.* But this is generally true of influential journals, even of those not devoted essentially to reviews. Five of the six most influential journals had 25 percent or more of their pieces in the form of book reviews. Only eight of the eighteen less influential journals had at least 25 percent of their items in the form of reviews and three of these were specialists in literary matters — the *Yale Review, Sewanee Review,* and *Kenyon Review.*

In the end, people, not topics or coverage, make a journal. The top intellectual journals attract more of the top intellectual talent. Let us go back to that list of the top seventy stars of the intellectual world to see how they fared in journal publication. The number of "votes" an intellectual received from his peers was closely related both to the sheer number of articles he had published and to the variety of journals in which he had appeared. The exceptions, of course, are the official gatekeepers themselves — the editors. They received a good number of "votes" but they tend to have their own by-lines appear less frequently than do other influentials. Most important, the top-rated journals are those with a greater proportion of works written by intellectual stars.

The specialized journals had more difficulty in attracting star writers. And it is specialization, of course, that makes a journal less widely influential as an intellectual medium. The three journals specializing in literature had the most difficulty in attracting the stars. In *Kenyon Review, Sewanee Review* and *Yale Review* less than 3 percent of the articles were written by stars. Specialized political journals do not fare much better. Of the five journals in which at least 40 percent of the content is political, three journals (*Nation, Progressive* and *Reporter*) each had less than 3 percent of their articles written by stars. But political journals with higher ratings of influence do attract more stars. Ten percent of *Dissent*'s articles and 8 percent of the articles in the *New Republic* were written by stars. This is an especially good showing for the *New Republic,* which is published weekly, features many articles per year, and must have a roster of people who can write regularly and promptly — not always a characteristic of intellectual stars.

As we said, book reviewing often serves the function of salon critique and discussion. So, to be influential, a journal's reviewers must have high prestige in the community of intellectuals. Of journals containing a large number of book reviews, the *New York Review of Books,* the highest-rated journal, had an outstanding record of 12 percent of its books reviewed by stars. Even the *New Yorker,* which does not specialize in book reviews but which received a high rating, attracted a high proportion of star reviewers – 15 percent. And it is our impression that the new editor of the *New York Times Book Review* must be assigning reviews with our list of stars in hand, for there has been a marked change in the status of reviewers over the last few years. Important journals which nonetheless do not have top ratings, such as *Partisan Review, Ramparts* and *Progressive,* compensate for their lack of star reviewers by reviewing the books of stars.

The Power Game

Intellectuals, through their reading habits and writing preferences, make some journals more prestigious than others because, of course, they approve of the topics, style, and writers of a given journal. But once journals have attained positions of eminence, they have independent power to make or break the prestige of individual intellectuals. This power is exercised through the clique and star system, the ability to publish some people and not others, and the ability to select some ideas and not others. And as will be evident, the power to support one idea while ignoring or denigrating another gives one the key to the kingdom of the intellectuals.

The paucity of stars, by definition, leads journals to an uncomfortable position. Most journals would like to get stars to write for them. As one editor put it,

> There is a difference between being important and being good. If I knew who the important people were, I would get them to come to [my journal].

The very search for stars means that journals which are most admired for their influence are also most hated for their cliquishness. The *New York Review of Books,* which heads the list of influential journals, also heads the list of journals receiving the most criticism about cliquishness, a distinction it shares with other top journals.

The *New Republic* was written by people who were talking to themselves. This also holds true for the *New York Review of Books.*

The *New York Review* is parochial, narrow, limited, and incestuous. They just review each other's books, take in each other's washing, and spread gossip about each other.

The other side of parochialism is power in the intellectual world. Only two persons we interviewed thought that cliques surrounding journals, a matter to be discussed in the next chapter, were powerless and only one person claimed that power was well diffused. When asked who had the power to make or break an intellectual's reputation, 45 persons said that journals rather than individuals held this power. Whereas some perceived the power to be invested in the groups or circles which center around journals – a matter we shall take up in the next chapter – others attributed the power to the editors, the reviewers, or the journal itself. The journal perceived as most powerful in this respect was the *New York Review of Books,* which was believed by 39 intellectuals to have the power to make or break reputations. *Commentary* was perceived by 19 persons as having this power and the *Times Book Review,* by 11 persons. *Harper's, Atlantic,* and *Partisan Review,* each selected as powerful by at least 5 persons, were selected primarily by noninfluential intellectuals. The only other journals selected as powerful by more than one person were *Public Interest* (selected by 4 influential intellectuals) and *Dissent.*

A few intellectuals perceived this power with annoyance. As one stated:

Power is in the circles around *Commentary, Dissent,* the *New York Review of Books,* and *Partisan.* Ninety-nine percent of what goes on in these circles is bullshit.

One felt the power of the influential journals was unjustified:

The people who control a lot of access are the people on the *New York Review of Books.* The most influential of these, I suppose, is Elizabeth Hardwick. If there is anything like a cultural establishment, it must be centered around that magazine. She [Hardwick] is also associated with the Theater for Ideas, which is an extremely important thing among intellectuals. It doesn't have much theater, any more than the *New York Review* has much to do with books. People say it ought to be called the *New York Review of Vietnam.*

A few persons were bitter about the power of the reviewers:

The *New York Review of Books* represents New York City itself and the attitude of *New York Review* towards books is very tough and very provincial. The *New York Review* is against novels.

The *New York Times Book Review* has a dangerous and threatening monopoly and has the only outlet for book reviews. . . . Unless your book is reviewed in the *New York Times* it won't sell.

For the most part, however, intellectuals did *not* make evaluative judgments or have negative reactions to the power of journals. Thus, although journals were criticized for their political positions, for being ingrown or incestuous, and for a variety of other reasons, the power of the journals was taken as a matter of course by most intellectuals.

As might be expected, the persons most often named as having the power to make or break reputations were the editors of the key journals – Robert Silvers, Jason Epstein,* and Norman Podhoretz. A few persons commented on the alliance between journals and book publishers represented by Silvers and Jason Epstein. Only two persons who were mentioned more than twice as having the power to make or break reputations were not connected with journals. However, one of these persons, Edmund Wilson, was said not to use this power. The other, Lionel Trilling, was highly regarded and respected by several intellectuals for his efforts in assisting talented writers to get published in influential journals. Although a few intellectuals named persons, academics in general, or publishing houses as being powerful, the control of access and the power to make or break reputations was primarily attributed to journals.

Although we have spoken of the influential and star intellec-

* Barbara, not Jason, Epstein is the editor. But many respondents, either by accident or by design, said "Jason," rather than "Barbara."

tuals as "making" a journal, the power of journals can obviously "make" an intellectual. According to our respondents, there are a variety of ways to get an "in," though there is no certainty that any of these methods will work. Some say that once a novice has appeared in any one of a number of reputable journals, he has it made:

The world of publishing is very small. Once you've broken through, with one or two pieces, then there are so few editors, and they all know each other, and they're so unimaginative that they keep you publishing the same things.

I think that if you get a long piece of criticism, which is the way to start, published in the *New York Review of Books* or get noticed by someone like Izzy Stone or publish a long piece in *New Yorker* or in *Harper's* or perhaps *Atlantic*, it's as good a way as any for setting up your reputation, because those are the reputable journals.

Apparently, however, it does little good to be published if the article does not appear in a highly influential journal:

X is an up-and-coming intellectual and he is going to be a very good writer, but nobody knows about him because he only writes for *Dissent*.

X has greatly influenced my thinking but people haven't heard of him because his magazine articles tend to be in general not in intellectual places . . . [he writes in] the *American Scholar* or something.

In contrast to those who advocate getting one article published, some said that the way to get published and get "in" is by knowing the "right" people:

I think a lot of what gets published gets published because somebody knows somebody who knows somebody.

However, others claim that "who knows who" may not make any difference:

I know Epstein and Silvers, but frankly, I don't think they would publish me because I am too middlebrow and I'm not radical enough.

> I can't get published in *New York Review* and *Commentary.* I know the editors but am resistant to getting involved with them, though not due to any lack of respect. I considered doing a piece for *New York Review,* but they told me, "No, you're not a serious man."

Despite this obvious gatekeeping function, the number of new names with presumably new ideas is not overwhelming. The *New York Review of Books,* which is granted by many the power to make an individual's reputation, began, in fact, by publishing those who were already influential. As a group, the older intellectual elite we sampled had little awareness of persons who were up-and-coming in the intellectual world. Many were unable to think of any up-and-coming persons at all, while others named such "newcomers" as Norman Mailer and John Updike. The editors were no more informed than the rest of the intellectual elite, for only half gave any names at all, and then few of these mentioned real newcomers. Most editors either did not know who might be up-and-coming or expressed considerable hesitation. One said regretfully:

> Oddly enough, the last major person to arrive on the intellectual scene with an unquestionable ability to write on both cultural and moral-political matters was Susan Sontag. I don't follow all of the younger writers and maybe there is someone emerging who is very good, but no one who has impressed himself or herself on my consciousness.

Not only do journals have the power to publish newcomers, should they choose to, but they may also reject established intellectuals who do not follow the current political line of a journal. Three intellectuals told us they were excluded from the *Times* and one from *Saturday Review* because of views which were unpopular or too radical. As one related:

> The *Times* once asked me to write a piece which I did write and which they refused to print . . . because it so shocked them. It was the first time that I had found myself censored, so to speak. I am not a revolutionary by any means. I was amazed to find first that there was an organization like the *New York Times* [*Magazine*] that still

held a fundamentalist opinion and secondly, that I would be excluded. I mean, by God, there were certain things you couldn't print.

Conservative intellectuals also had difficulty in getting printed. After one liberal intellectual commented that he could not get published in the *New York Review of Books* because of political differences, he was asked if there were other journals where he could get a hearing. He replied:

About the Vietnam War? How would I, today, in 1970, go about getting articles about the war published in order to reach other intellectuals with my position, which is not a popular one? There are three places where I think that my position might get a hearing. I know that it would get a sympathetic hearing from *Commentary* if I wrote it freshly enough so that it had something to offer anybody. I imagine the *New York Times Magazine* section could be persuaded and, I think, *Harper's*.

A conservative, who has been widely known for his hawk views on Vietnam, has tried these places and finds it impossible to get his views printed or reviewed:

[My last book on foreign policy] was really blackballed by the main instruments of book reviewing. This was not by accident. The first time in my life that I've had a book not reviewed by the *New York Times*. At the moment, the intellectual media does not feel like publishing me, but I'll keep trying. I can get a hearing through speaking but at the present time I can't get published because of the blank wall of resistance against the views of foreign policy that I represent. I feel, sometimes, almost totally boycotted by the effete snobs, as they say . . . *Harper's*, *Atlantic*, *Commentary*, and the *New York Review of Books*. I know all the editors and can call them and they will listen, but I can't get published. This view or this group of views are not being reviewed, they are not being discussed, and they are not being printed and editors say flatly they just won't do it. We are in the grip of a situation in which the open market for ideas is not very open.

Perhaps at one time there was an open market for ideas, but we doubt it. The function of a journal just like that of a salon is, after all, to select from the many new and old ideas those which it feels should now be talked about. Nearly all respondents emphasized that the influential journal had a definite political stance, mainly of a leftist nature. The *New York Review* was perceived as the most radical of the influential journals, and most respondents did not like this. The famous Molotov cocktail cover came in for particularly negative comment.

> *New York Review* devoted a cover on how to make a Molotov cocktail. I think this fashionable emphasis on violence comes from people who don't have very much experience with violence.

> My crowd was outraged by Kopkind's articles, or morality out of the power of a gun . . . at the complete going overboard in favor of the riots, at the Molotov cocktail on the cover, etc.[6]

The curious thing is that most elite intellectuals we sampled were opposed to radical politics but believed that the intellectual community was influenced by these politics. In other words, most of the elite intellectuals we talked with are not influenced by the alleged radical *New York Review* line and do not agree with radical politics, but perhaps because it appears in print they assume that other intellectuals are influenced by this line.[7]

Because journals do have a line many complained that the journals stifled and rejected novel ideas. Some felt that journals were ingrown and perpetuated the same repetitive line to their small and agreeing audience. Influential journals were accused of being followers rather than leaders and of summarizing what everyone already knows rather than entertaining original ideas. That is, journals were seen as behaving like universities, king-making institutions and other more rigid gatekeeping organizations, rather than as flexible salons encouraging the exchange of ideas. This feeling was especially apparent in the comments of intellectuals whose unpopular or novel views were not published or reviewed. One intellectual complained that none of the influential journals would accept his piece opposing the Warren Report because at the time he wrote, opposition was novel and unpopular. Later, however, all the journals were looking for articles on the assassination and his was eventually printed. The failure of journals to originate topics is expressed in a variety of ways, for example:

> The *New York Review of Books,* you might say, in one way or another has replaced the *Partisan Review* as intellectual satire, but the *New York Review* doesn't really originate anything.
>
> I must say that I have not found any particular views brought out in recent years by any of these people who contribute to these magazines [*New York Review, Commentary, Partisan,* and *Dissent*] that could be said to be more than somewhat provocative, and by and large, extending traditional forms of thought pro and con. The attempts on the part of the New Left to introduce some new ideas about politics have been a total failure, largely through lack of serious consideration.

To be sure, new ideas must eventually enter the journals, but they do so in a special way. New ideas are contributed mainly by new people, that is, not by the established intellectual stars, who are almost all generalists, but by long-term specialists in the current hot topic. An editor wishing to cover an issue which suddenly becomes hot looks to writers or to academics who specialize in it or they send out a staff reporter to cover the scene. Of the articles covering political topics which became hot in the middle sixties, 60 percent were written by academics and most of the remainder were written by reporters or staff writers. And this is as true of campus crisis issues (which one might expect to be covered mainly by academics) as it is of the war in Vietnam. The most influential writer on Vietnam was an academic, the late Bernard Fall. Also influential were reporters such as Shaplen, Halberstam and others who specialized in Southeast Asia. An early antiwar stance was taken by Hans Morgenthau, an academic and a star who had already written widely on foreign policy and the Cold War, the issue of most concern to intellectuals since 1945.[8] Influential intellectual generalists such as Theodore Draper, John Kenneth Galbraith, Mary McCarthy, Arthur Schlesinger, Jr., Susan Sontag, and others did not take a public stance against the war in major journals until most intellectuals had already turned against the war.

This brings us back to the power of journals to make people.

The journals turned to the Vietnam specialists for help. In turn, some Vietnam specialists became highly influential because of their appearance in these journals. Chomsky is an interesting case of an academic who had been a technical specialist in a field other than foreign policy but who was catapulted by exposure in the *New York Review* on the topic of Vietnam into becoming a more general intellectual star. Indeed, he has made a point of urging other academics to abandon narrow specialization in favor of general political and moral interest. In his case, the rewards were in terms not only of a feeling of moral rectitude but of prominence in the general intellectual community. One can cite someone such as David Halberstam, who was unknown to most intellectuals before his own Vietnam reporting.

Finally, even if an intellectual has the right political line, the proper expertise, and the right connections, he must adopt an appropriate intellectual journal style or he cannot get published. An acid tone is said to be helpful. One faithful *New York Review* reader claimed that the people who write for it "are as hate-filled as the thing they are attacking." A number of persons felt that the *New York Review* was caustic in its criticism of books and "if they don't knife the book, they knife the author." And one former *New York Review* critic claims he no longer writes for them because "I don't want to do an assassination job." On the other hand, sheer "good writing" is held to be of great importance. One respondent, noting that the *New York Review* tends to be regarded as elitist, explained that it is "partly because it is awfully hard to find writers for the damn thing. You need people who combine expertise and writing skill — very easy to find either and not so easy to find both."

Whither the Journals?

Influential intellectuals can make a journal and journals can make or break the reputation or prestige of intellectuals. Changes in the times also affect the influence of intellectuals and of journals. Since the middle of the Johnson administration, intellectuals have been out of favor politically. The Nixon administration contributed to economic conditions detrimental to many journals. It not only failed to consider the views of intellectuals but the Vice President went so far as to ridicule the intelligentsia.

In the past twenty-five years there have been a number of shifts in journal influence. The *Nation* lost prestige during the Cold War period, while the *Partisan Review* and *Commentary* gained. Then, as the trend shifted to the left, the *Partisan Review* lost prestige; the *Reporter*, partly for political and partly for personal-financial reasons, had to cease publication altogether; and the *Kenyon Review* also stopped publication partly because its emphasis on literature was less appealing. The *New York Review* began publication and rode the crest of the new wave of antiwar radicalism, while *Public Interest* took an opposite tack and covered the neglected field of public policy analysis at a level of general discourse.

At the onset of the seventies a number of developments have made the future of intellectual journals and their relative influence and prestige even more uncertain. To begin with, the mass-circulation magazines such as the *Saturday Evening Post* (already defunct in 1969) and *Look* went out of business. *Life* joined them in 1972. In addition to revenue and circulation problems, all journals are threatened with vastly increased postal rates. The general economic uncertainties have further exacerbated the weak financial position of journals. Those intellectual journals which aimed at a relatively wide audience, such as *Harper's* and *Saturday Review*, have been especially hard hit.* New managements have attempted to alter the aim of these journals away from the more intellectually prestigious but perhaps more closed circle of penumbra intellectuals which editors Willie Morris and Norman Cousins targeted.

The case of the *Saturday Review* is especially instructive in view of its financial failure and the return of Cousins to command. Basically, what the new management of *Saturday Review* tried to do was *segment SR* so that it would appeal

* To date, analysis of the audience of various journals has been confined to marketing research reports. All these journals have well-educated readers – mostly college level or beyond – with impressive incomes, professions and levels of political participation and interest.

more to specialized audiences in the manner of their previously successful *Psychology Today*. Even while they were designing this change we went on record (alas, in mimeograph form only!) predicting failure. The reason for their failure could have been deduced from our analysis of how to publish a successful intellectual journal: it *must* be general and must cover a very *wide range* of topics. While specialized journals have in recent years proven more successful on the semi-mass market, the intellectual journal itself is a specialized form — except that its specialty is to be brilliantly general, capable of immediate comment on whatever is the "hot" topic of the moment.

In line with what seems to have been the trend of the early 1970's, journals deemphasized politics, perhaps because the liberals and radicals who make up almost all of the elite intellectual journals' audience felt so frustrated in this period. The Nixon-Agnew team seemed for a while to have effectively defused the "movement," and political issues took on a more pragmatic or technological air rather than the ideological or moral flavor of some of the commentary of the sixties. Watergate, of course, has put this shift into question, and concern with the evil nature of the "power structure" may again be of interest to intellectuals. Whatever the case, although I. F. Stone joined the *New York Review of Books* (and turns out to be the favorite author of its readership), the *Review* does seem in recent years to have a less radical air about it. It has even published Buttinger on Vietnam, a man who was once a strong supporter of Diem, and regularly publishes reviews by Passell and Ross, whom Jason Epstein himself faintly damned in the pages of the *New York Review of Books* as "liberal economists." *Commentary* seems definitely to have moved to the right, but its swing began in the late sixties when its editor became incensed (along with most intellectuals — as we shall see — who did not, however, own up to this in print) by the stand of the New Left.* The move of *Commentary* along with the growing influence of the new journal, *Public Interest*, has led to the coining of a phrase, "The New Conservatives." In an attack upon them in *Dissent*, Joseph Epstein (who himself has appeared in *Com-*

* Podhoretz, by the way, does *not* himself feel he or his journal has moved to the right.

mentary) notes "how various a lot these men are, as well as how very different they are from those whom till now one had been accustomed to think of as American conservatives. Many of the new conservatives disavow the label,"[9] and – we might add – those so labeled whom we interviewed in 1970 still showed strong evidence of their recent liberal-radical past.

Finally, wherever the journals are going, they must in some way take into account the changing tastes of the newly arrived well-educated youth audience. While it is true that the *New York Review of Books* has been well established on college campuses, perhaps more so than any other intellectual journal, no major journal (except perhaps the *Village Voice*) has been successful in regularly attracting contributions from writers under thirty.

Intellectual journals have been important in the United States because they have captured and communicated the intellectual currents of their time. But the journals and times are changing so rapidly that some of our statements may be outdated by the time this appears. In an age of great intellectual change and uncertainty it will take a remarkably creative effort to present a coherent, meaningful representation of the intellectual community, if such a community still exists.

3

The Leading Circles

Elite American intellectual life consists of a loosely allied and interpenetrating trinity: the prestigious intellectuals, the prestigious journals, and the leading circles. Leading intellectuals write for the leading journals and the combination of journals and intellectuals produces the leading circles. Each acts to bolster the other. A journal is important because people who are already important write for it; a person becomes important because he writes for an important journal; and the social network assures that only important people write for important journals. In this chapter, the last and most crucial member of the "three" — the leading circles — will be dissected.

The picture of an integrated system of stars, journals and circles exaggerates present reality. While most of the leading intellectuals are members of the leading circles there are important exceptions; some significant circles contain insignificant members; and while some journals have a coterie of contributors, the relation between journals and intellectuals is looser than it may once have been. Not only are replacements for the current generation of leading intellectuals difficult to find, and not only is it hard to predict which will be the leading intellectual journals of the next decade, but the very warp and woof of the fabric of intellectual interaction has been seriously weakened. The circles still exist; they are important; they do contain important people, but they are not tightly linked nor are they solely reflective of New York or its intellectual journals.[1]

Intellectual circles can of course be based on many elements other than intellectual journals. Professional activities, ideological stands, academic organizations and societies all compete as possibilities. But of all the factors which cement relationships or at least make them possible, geographic propinquity is most obviously visible to the naked eye; neighborhood is the archetypical cause of nonfamilial human relationships. And so we observed in Chapter 1 that the geography of intellectual life was of great concern to the American intellectual elite. Their geography, however, is totally ambiguous since half live in the New York metropolitan area (our "lunch distance" definition) and half do not. An observer is therefore justified in saying that "*fully* half" the American intellectual elite live in New York, but another observer is equally correct if he notes that "*only* half" live in New York. For this reason and because matters are in flux, the members of the American intellectual elite themselves agree neither on the shape of their society nor on the nature of relations within it.

The way intellectuals see their own society is important, for ultimately the way they see things affects their actions. Moreover, their views are certainly no less relevant than the descriptions of entire societies and cultures which have been given to anthropologists by one or two village elders. Our "natives" are trained, expert observers many of whom have themselves written on the topic of intellectual circles. The high quality of our informants demands that we give considerable weight to their own opinions about the state of American intellectual circles even though their opinions are often contradictory.[2] In the first part of this chapter we shall therefore consider in greater detail than in Chapter 1 their views of the degree to which there is or is not a center to American intellectual life, the relative importance or lack of importance of New York, and the claim that the university is taking over, and that things in general are not what they used to be. In the second part of the chapter we shall go beyond what is visible to the naked eye and look at the social circles and networks as revealed by "who actually associates with whom." Some of the myths about the circles will be supported, but many will not.

The Way Intellectuals See Themselves

Whether there is any center at all to American intellectual life is questioned by some, while others believe that if there is a center it is not geographically defined. Yet only about 15 percent thought there was no center to intellectual life in the United States.

A novelist who delights in living in the countryside miles from New York offered allusions to other countries and other eras.

> Historically, I don't know of any other similar situation. The American writer is much more reclusive, much more isolated than any other writer in other literary communities in which I have lived, particularly Russia where the writers will meet at the Writer's Club in the evening and talk. American writers don't actually avoid one another—I always love seeing the writers that I know. But when we do meet—roughly about twice a year—there isn't any power complex involved and there isn't any social life involved. . . .
>
> Occasionally, a woman of indescribable ambition will try to gather together in New York a salon of seventeen or eighteen acceptable writers, but it never seems to work terribly well. . . . We don't have, and never have had in the United States anything that could be described as a literary community.

Several intellectuals specifically contrasted the absence of a center to the situation of Paris in France or London in England. Another felt that intellectuals are located in many areas of American life.

> I try to divide the community up into a series of pyramids. . . . You've got your intellectuals in all of these groups. The power of an intellectual in the academic community is quite a distinct thing from the man in public policy, and in the entertainment business it is a

wholly different kind of thing from the art business. Who are the important intellectuals? Well, they're all over the lot.

Finally, a prominent intellectual interested in psychological matters thought the whole idea of an intellectual as a separate entity makes no sense.

> I don't feel it that way because I don't feel that I am an intellectual. You know, I'm a person who likes to play handball and make love, and so forth, and I have a pretty good head. . . . I don't feel that "we" as intellectuals form a group.

The rest of the intellectuals who commented about intellectual centers and circles all agreed that there were some kinds of groupings, however diffuse.

Only a few, considering that intellectuals are "men of ideas," described circles in terms of ideas or ideologies; some linked ideas to journals and some to key individuals; others spoke in terms of occupational divisions with the university as a key factor. But most of those who talked about intellectual centers discussed geography and social interaction. Despite the fact that ideology, occupation and journal affiliation are in principle nongeographically bound, most intellectuals discussed them in the context of the state of affairs in New York City.

The Importance of New York

Some intellectuals talked only about New York, others found New York the main center although other centers were important too, while a third group felt that New York wasn't what it used to be. An expert on American intellectual history said simply, "It's [the center of American intellectual life] New York, and I think it is significant that it is not in the Capital. [Intellectuals] are cut off from the knowledge of political personnel and the problems of politics . . . in a way that would not be true . . . of English intellectuals." An intellectual who lived in Washington, D.C., concurred, "Here I am stuck down in Washington, D.C., and everybody tells me the power and culture is in New York." And a West Coaster, after having talked at length about his experiences in New York, was asked about his present location:

> No, I think the West Coast culture is very, very thin. . . . You know, there aren't very many first-rate writers in San Francisco. They're more likely to be provincial. I mean you get a certain kind of provincialism in New York but even more so in San Francisco. . . . I think it's a mistake for a writer to stay out here, and a mistake for a writer to keep on working in the movies [as he does], and a mistake for a writer to keep on teaching for too many years.

A New Yorker summed it up by saying, "New York is pretty much 'it' and the others are satellites."

The degree to which New York is the sole center, however, was debated by many. The majority thought that although New York was the center, colleges were oases in the wilderness. Others saw not one main center and a series of outposts, but a corridor between New York and Cambridge-Boston.

The Changing Scene

Some of the most interesting comments were made by the relatively large number of intellectuals, some residing in New York and some not, who felt that while New York was indeed an important center, its importance lay more in the past than in the present. Again, many compared New York with London and Paris and emphasized that, unlike Europe, American intellectuals had little contact with each other, even when New York was more important and more cohesive with ongoing political events. Moreover, even at best New York did not have the cohesive intellectual life of a European capital. An ex–New York writer felt nostalgic.

> To me it was very exciting to get into the *Partisan Review* group in 1946 because it was still alive and fresh and very vigorous. [Today] there isn't even a *Partisan Review* group in the old sense. . . . I doubt whether today there is a group in the sense that is socially organized and so on. . . . French literary life is much more cohesive.

Several intellectuals in the New York community denounced what appeared to them to be a decline in the fraternal quality of New York intellectual life.

You have this New York community which I think is somewhat ingrown, snobbish, elitist. And what impresses me about the intellectual world in New York today by contrast with the intellectual world here when I first knew it . . . in 1929, was that there was a generosity of attitude then and it continued all through the 30's – a receptivity, a general friendliness and so forth that seems to have gone by the door. There's more backbiting [today], vicious polemics, knifing, and whatnot. . . . I don't go to cocktail parties for books that are coming out and that sort of thing any more because it seems to be that the pastime is to stand around and stick stilettos in every name that's mentioned.

Some intellectuals implied that the acerbic tone was set by the entrance of Jewish intellectuals onto the scene, while others felt it was the result of McCarthyism.

A leading intellectual, who according to other persons' accounts was right in the heart of things and had been so for many years, did not feel that New York groups were cohesive, if they existed at all, and was quite ambivalent about the place of journals in the scheme of things. Since everyone acknowledges him as expert and because he has never written at any length on this topic, let us quote him at length.

We have a tendency to factionalize. In New York I am always amazed at how many people I keep running across all the time at parties, dinners, etc., that first of all I haven't met before and secondly that don't seem to be related to any groups, if there are any, that I might belong to. I would say that at present, anyway, I belong to the let's say sort of *New York Review* crowd, whatever that is. And then I suppose there is a *Partisan Review* and a *Commentary* crowd, etc. They seem to be crystallized around magazines, I would think. But I don't think that it is very significant, really. No, I would say that in New York, first of all, you have a couple of dozen, in my case, personal friends that you see every few months or so that you think of inviting to your place or they invite you. And those friends really have quite a different prominence. There isn't a question of a crowd or a clique. . . . You know, Norman Podhoretz in that *Making It* . . . doesn't really tell the truth. [He] had this idea of the New York family, etc. There was something in that if you take it over about twenty years, but certainly not if you bring it up to the present. It certainly cannot be thought of as a habitual-social-discussion-meeting-often kind of business. Whereas in London, I had the impression that this was the way there, that you kept running into the same people and that there was a certain continuity which I think is lacking here. I think Podhoretz's New York family is really a question of the people that write for the same magazines, that read each other's stuff and in that sense are connected, but don't really have so much to do with each other in this way.

Those respondents who felt there was a strong integrated New York intellectual circle even today have been in fact out of direct touch with New York for a while. Another historian who is an acknowledged authority about American intellectual circles but who resides some distance from New York, now says, "When I go to New York I feel very much an outlander [even though some key members of New York circles are friends of his]. For example, I've not been to New York all this year, but I've been to Washington many times." He reports about New York that there is

sort of a Mafia. . . . They gang up on you if you're an outsider. . . . I used to go to Jason Epstein's parties when I was down in New York, and the same people were there all the time – Dwight Macdonald and Norman Mailer and all the rest of them. And they sort of defended one another, you see. . . .

One reason for the diversity of opinion not only about intellectual circles in general but their character in New York is that things are indeed changing rapidly. A historian explained:

You had a former grouping but it fell apart and now you are falling into a period where . . . that [circles and their break-up] happened some fifteen years ago so you have a great deal of fragmentation and you have to deal with people's memories. . . . There are times when you do get circles and you do get alliances and there are times when they break apart and new ones form. . . . So the problem might be what happens at any particular time, but also, what explains the formation and disintegration and then the reformation of the groupings. If you don't take an historical view you tend to get really outlandish deductions from very immediate situations.

While we cannot here present a history of New York circles, both the reminiscences as well as the biases of some of our respondents will help to explain what is happening so that we can avoid "outlandish deductions."

To understand the changing character of American intellectual circles both in New York and elsewhere, we have to see what foci there are for such life. Thus far, some of the top journals have been mentioned as well as some intellectual stars. Fashions in both have been changing over the last ten years, as we noted in Chapter 2. Moreover, the *New York Review of Books, Partisan Review* and *Commentary* are hardly the only pegs upon which circles hang.

A free-lance writer who writes mainly for the *New Yorker* had this to say:

> You want me to give you some phony sociological analysis? There is a little self-regarding group that is sometimes called the Establishment, that consists of the *New York Review of Books,* the *Partisan Review* and circles around them as opposed to groups like the P.E.N. Club which is more of a middlebrow writers' organization. . . . There is this group that I spoke of, the Epsteins, Robert Silver, Dwight Macdonald, and so on.
>
> Q. And they are the *New York Review of Books* and *Partisan Review* group?
>
> A. Yes, and Norman Podhoretz, and so on. . . . They are the reigning literary group. . . . I much more identify with the middlebrow – well, *New Yorker* used to be pretty much of a group by itself, and it is to some extent now. The *New Yorker* writers only see other *New Yorker* writers all the time and it used to be sort of a movement, but it isn't so much so anymore.

A few intellectuals mentioned artists and art galleries as centers, but almost all were in the position of the writer who said, "I could talk about the art world a little bit, but I don't know much about it." Some few spoke of newspapers, but almost no one mentioned the theater, or music as possible pegs. One man's circle hung about WBAI. A younger professor of

English at a leading New York City university, while conceding that the "New York intellectual-cultural-literary group is *sui generis* . . . and is traditionally the most striking intellectual group in the United States," was willing to include within it far more than a limited set of top intellectual journal star authors. "The New York intellectual group includes publishers, magazines, literary people, editors, writers and academics in the New York world as well as some people from high society, swingers of various kinds and from some of the other arts apart from literature." Such catholic views are in the minority. Though the world may not begin and end with the *New York Review of Books* and a few other journals, by far the majority of intellectuals we interviewed tend *not* to include publishing, music, theater, movies, TV or the plastic arts, and certainly not "society" and "swingers."

Which brings us to the matter of "hangouts." One reason European capitals have a greater unity of intellectuals within their own ranks, as well as an extended network of intellectuals in the arts, theater and movies, is that they all tend to be together in a few cafes or clubs. Such gathering points seem less effective in New York. Let us return to the interview quoted above with a leading expert on New York circles.

Sometime in the late thirties, it occurred to a number of us that it would be a good idea to have a place in New York – one place – like the many cafes in Paris, where between five and seven if you drop in you will see people that you know and can talk to without making an appointment. And there wasn't a single place then, so anyway, we tried to make a deal with some Greenwich Village bar, I think it was "Julius" or something like that. That was hard enough because they weren't so sure that they wanted to set aside a room, but anyway the main trouble was that actually people didn't come around – there was something artificial about it – we actually did see each other anyway. So that didn't come to anything.

Q. Anything today?

A. Well, that "White Horse" thing closed down and now there is something called "The Lion's Head." I was there once with X and I must say, there was quite a lively crowd of mostly journalists, like Pete Hamill and people like that, that guy from the Village Gate, Art D'Lugoff . . . and also I gathered that regulars went there practically every night. Well it has to be in the Village – I can't think of any other place in the City where . . .

Q. How about "Elaine's"?

A. Oh, "Elaine's," I've been there several times, but that's a dinner place that is not quite the same as a place to drop in; you have to have a dinner, you know. Yeah, "Elaine's" certainly was the place to go but I don't know what it is like now. But it's odd about New York, and again, it's the same point I was making about the fragmentation of the cultural-intellectual life here as against London, and I'm sure against Paris, that there aren't any of those places, or at least very few of them.

Q. And there's no salon or anything like that?

A. Well, maybe X still has it, I don't know, he hasn't invited me for years. I don't know why he did it, but he had a rather large place on Riverside Drive and quite frequently, about every two months or so, he used to get together about sixty or seventy people, sort of the liberal and left liberal establishment – people like Jimmy Wechsler, the more enlightened trade union people, people like I. F. Stone – and it was sort of more political than anything else. But it's true he did do that for years, but I haven't heard of anything since. You'd think that some rich, idle . . . some women that had interests in the arts, letters, politics, would then do it, but as far as I know they don't.

The gap noted by this observer has in fact recently been filled by several fairly active salons – but these do remain by their very nature limited to some intellectuals while excluding others.

One reason why there is no central hangout or sets of hangouts in New York is that New York intellectuals rarely seem to agree upon anything. Take the restaurant "Elaine's," which our "leading expert" put down. Another man about town thinks it is very important.

There's a restaurant uptown called "Elaine's." . . . You almost automatically seem to go up there for supper. And one gets that vague feeling about "Elaine's" that if you dropped a bomb on ["Elaine's"] you'd wipe out *Esquire, Paris Review,* to a lesser degree the *New York Review of Books,* a couple of publishing houses – Random House; wipe out a couple of theaters. *Scanlan's Review,* that would go down . . . *Harper's,* a lot of *Harper's* people there. You'd put quite a dent in the community, I'm sure.

Another writer talks about "Elaine's" and about Mailer's circle, though with less enthusiasm.

> You can see Mailer just victimized by his little circle in New York. . . . "Elaine's" in New York, where writers go . . . is full of fops. There are some good-minded writers who come into "Elaine's," but anybody who hangs out there for very long can't be a major writer. . . .

Which is perhaps why the first man talked about bombing the place.

One reason, of course, for the demise of hangouts, if indeed they were ever effective centers, is that the New York intellectual community is growing older and there seem, at the moment, to be few replacements in their ranks. Even the man-about-town, who is one of the younger intellectuals, is not up to long bouts at cafes anymore.

> The "White Horse" used to be popular about ten, fifteen years ago. I used to go down there all the time. . . . I don't know what goes on there anymore. There's a whole thing in the Village now that I don't know anything about . . . I get down there so damn rarely. You find that there's less and less of the café scene going on at least as it was in the old days. You get tired. Get home too late.[3]

Jewishness as a possible focus for circles seems to be latent in the remarks of many we interviewed. Some brought it right out to the surface, though always with a touch of ambivalence. A New York Jewish intellectual, when asked where the center of intellectual life in America was, replied:

> Oh sure, the New York Jewish intellectual milieu, no question. [There are] honorary Jews who aren't Jewish. . . . There is something about the Jewish intellectual critical traditions that has formed an important core in this city. . . . If you run down the list of the leading hundred intellectuals, at least, I do believe you would find an astonishing proportion of Jews. . . . And yet it's very surprising. I'm not at all a sort of proselytizing Jew, but there has been a long history of Jewish intellectual endeavor. . . . It was partly in Berlin, it was partly in London, partly in Paris, and always quite powerful here. Odd thing that this would be the case.

The alleged Jewishness of New York intellectual circles is usually qualified by respondents who realize that many leading intellectuals are not Jewish, as in the reference to "honorary Jews," and references to "Lowell, and Mary McCarthy, and Edmund Wilson — you think of Wilson, although he lived in New England a good part of his life, you think of him as being that sort of New York intellectual. . . ." But the most important qualification of the Jewishness of New York intellectual life is that the Jewishness itself is qualified.

Before the war none of us was Jewish in this way [deep-seated loyalties]. It was the Hitler War that made us so. We were much more international. There was a great belief that socialism and modern culture would somehow eliminate all these provincial things. And in a sense the religion of the Word and the religion of Socialism would take care of the religion of the Jews.

What happened to the New York partly-Jewish leading intellectual crowd of writers for the chief intellectual journals? Why, according to so many of our respondents, are they not now as unified or as important as they once had been? The decline of New York as a place to live, the concomitant rise of a system of regional universities with important resident intellectuals and the breakup of the Old Left, first as a result of McCarthyism and second, and more crucially, as a result of the youthful wave of the New Left and the rise of an antirationality movement, all seem to have contributed to a decline in the vigor of the New York intellectual community.

An important critic, for years associated with New York and who still lives there and who could say, as many did, "Sidney Hook was one of the first people to influence me politically in 1931. He made a Marxist out of me in order for me then to throw it out . . . ," gives us a capsule history of the alleged decline of New York.

[New York] has been [the center] until very recently. The important literary journals were edited in New York, and the people who wrote for them were largely living around in New York. The *Nation*, the *New Republic* when it was younger, *Partisan Review*, *Commentary* in the forties and fifties, and the *New York Review of Books* in the sixties — it was no accident, it was really pivotal. Those were the pivotal journals of the intellectual world, I think, and . . . I don't mean that only New Yorkers wrote for them, but so many did that this made a community.

New York is scattering, the intellectual community is breaking up now. We're going back to an earlier phenomenon in America – the isolated writer, the intellectual or the tiny little cadre, a little group here, a little group there. It's not what it used to be. I suppose its cause, if I'm correct in saying this is happening, is the deterioration of New York as a city and the general fleeing from New York that is [consequently] taking place among intellectuals. Some of them are very glad to go somewhere else. In the last two or three years I've met an awful lot of people who have left Columbia, left New York altogether and gone to other colleges . . . and you know, people are not so eager to live in New York if they weren't born here.

Why Circles Might Be Weakened

The post–World War II growth of American universities and their willingness to admit Jews to important positions, together with the decline of New York City as a place to live, have certainly contributed to the decline of New York as a center of intellectual life. This very decline may in itself weaken the strength of whatever direct interaction still takes place among New York intellectuals because many key members of any given circle do not now live in New York. But, as we said in Chapter 1, the persistent envy of American intellectuals for Europe and their consequent use of European capitals as model intellectual centers[4] disguises the fact that the United States may never have had the centralized intellectual salon life characteristic of Berlin, Paris, or London. When all is said and done, however, we are inclined to agree that the circle and network activity of American intellectuals may be weaker than it was in the period between the two World Wars or immediately after World War II. Indeed, the most important reason for the apparent decline of New York is tied to the very basis for the ascendance of the present generation of leaders in the first place: politics. That "old gang of mine" held together, for the most part, in its newly found nationalism and liberalism in the fifties; but the sixties tore it asunder.

An important participant felt that the threads which held the New York intellectual community together were, in the thirties and forties, Jews and "the radical movement":

> You see, it was at the time when *Partisan Review* was almost the very Bible. *Partisan Review* satisfied all the criteria: it was Left, it had been in the Communist Party, it broke with the Communist Party and then especially when it broke with communism itself and became sort of an organism of the new libertarian impulse. Many of us viewed it as *the* magazine. It had very high standards and it had this peculiar combination which was terribly important of the greatest possible receptivity to avant-garde literature and painting, along with radical politics. . . .
>
> *Politics* was founded because Dwight [Macdonald] felt that *P.R.* wasn't sufficiently political, and since then all the magazines, like *Commentary*, have had some of this function. . . .
>
> The problem is that all of us that were left-wing in the thirties inevitably have gone in different ways since then, not just because we have become the establishment, which we definitely have — it seems incredible but we have — but largely because of two things: one was the natural cautiousness that comes at a certain age . . . and the other has to do with . . . the great fear of totalitarianism.
>
> One of the characteristics of intellectuals today is that the young ones don't care a hang about history and the left middle-aged ones like myself would like to think of themselves as still being left of center but inevitably do compromise because they feel that with so much of the communist world against us that the situation is obviously tragic. Well, all this points out the contradictions, and you might say, the deep-seated loyalties that are involved.

In short, the history of the recent past seems relevant. The old radicals still have their loyalties, but they are hard put to say to what they are. Whatever the case, they feel uncomfortable about the young. As we shall see in Chapter 6, almost one fourth of our sample discussed the problem of campuses, youth, the "movement," and the New Left at great length. Most in our sample who discussed the topic were essentially negative.

The sixties intertwined politics with a style which was repugnant to much of our sample and the stylistics as much as the ideology of the New Left or the "cultural revolution" helped to break apart the old circles. A critic in our sample who is an advocate of the new style explained the split in this fashion:

I think there is a division between those who want to protect certain traditions – intellectual and literary traditions . . . and those like myself who feel that . . . these traditions have led to disastrous inflations of the importance of "literature" and a very damaging increase in the value of unnecessary terms [such as] . . . "Humanism" or "Reason" or "Rationality." The argument is between those who want to perpetuate traditions [such as] Arnold and other rational philosophy – and the jobs . . . ranks . . . and hierarchies that go with them – and those who don't. Those who don't, like myself, are very often called "swingers" by those who do. . . . The kinds of people I am talking about would be in some of the eastern universities . . . like Lionel Trilling. . . . These people are against the study of popular culture.

Most of the published comments about breakup and controversy in the intellectual world today have centered about intellectual journals.[5] Our respondents, who talked about these matters before they became more commonly written about, tended to mention journal and political alliances in the same breath. Here are some of their views of the current scene.

The New York intellectual community . . . split up mainly over political differences, especially in the sixties. Silvers has replaced Howe on the left, while the *Public Interest* crowd are technocratic and antiradical people. Many New Left intellectuals have a guilty conscience because they were quiet during the 1950's. *Dissent, Commentary,* and *Harper's* belong to the Democratic Left.

The *New York Review of Books* group is eclectic, including Old Left, radical young writers of the New Left, persons influential in book publishing, and English literary critics.

There is the group of Old Left like *Dissent* and *Partisan Review. New York Review* is more eclectic with the New Left, Old Left, academics, and English.

Political differences do not come about without symbolic events. The Vietnam War in foreign policy and the New York City schoolteachers' strikes of 1967–1968 in domestic affairs may well have been the key symbolic events of the sixties. The war is an obvious issue, and one to which we devote a good part of this book. But in many ways the New York intellectual was more affected by the teachers' strike. Why should such a local, relatively trivial matter (in a national or international perspective) have been so important? The basic facts are complex and beyond the scope of this book.[6]

On the surface of it, the teachers struck because schools in the predominantly Black Ocean Hill–Brownsville district of Brooklyn, an experiment in community control partially supported by a grant from the Ford Foundation, refused to allow those teachers whom they believed ineffective with Black children to continue on the staff of their schools. One of the signs of this ineffectiveness, they claimed, was the fact that the teachers had struck over higher wages and better conditions one year previously, leaving many Black children without a means of education.

What is important is that the issues were differently perceived and interpreted by the intellectual elite in general and by writers in the *New York Review of Books* and *Commentary* in particular. For the strike was one of those rare occasions in which a set of issues that had once come together as a package was totally unraveled, forcing an entire new set of alignments. At one time, the New York intellectual elite (and by some extension, the rest of the American intellectual elite) thought these truths to be self-evident axioms of their creed: unions, especially teachers' unions, represented a progressive force and all progressive forces deserved some support; the education of Negroes (or Blacks, as they came increasingly to be called in the sixties) was of poor quality and its improvement was essential if Negroes were to achieve their proper place in the United States; all peoples had a right to self-determination; militant civil-rights activities were necessary to advance the Blacks; anti-Semitism, associated with fascism and Stalinism, was bad in itself and doubly bad because of these associations; modern-day bureaucracies were a menace to desirable humanistic individualism; rigid forms of education and rote learning were evil. In addition, an important segment of the intellectual community was coming to believe that decentralization was an answer to the impending crisis of *1984*-style "big brotherism." The genius of the protagonists – led by Jewish Albert Shanker, the United Federation of Teachers Union president on the one hand, and Black Rhody McCoy and CORE leader on the other, was that they managed to structure the situation in such a way that the attainment of some of these values seemed to negate the attainment of others.

While there were enough issues involved to create not two sides but half a dozen, the characteristics of a strike forced polarization. Many socialists, Old Leftists, and others associated with such journals as *Dissent* and the *New Leader* felt that union control over hiring and firing was the most important issue and that anyone who was truly progressive ought to see this. In this view they were joined by some Black leaders such as Bayard Rustin and A. Philip Randolph. Writers for *Commentary* magazine felt that the issues were really Black anti-Semitism, aided and abetted by the WASP establishment, and so they supported the union. (Nathan Glazer, writing in *Commentary* one year later, however, pointed out the large number of Black Panther supporters who were Jews.) In addition, some felt that middle-class Jewish teachers who had "made it" were being deprived of their rightful gains. Writers for the *New York Review of Books* and their friends, who felt "radical" rather than "liberal," favored community control, disliked the New York City educational bureaucracy, and felt that entrenched teachers were using old methods to cope with new problems, and so supported Ocean Hill–Brownsville and castigated those who did not. An intellectual who had come down squarely on the side of the Blacks and the community organizers complained about *Commentary:*

> On the whole Ocean Hill–Brownsville community control school issue they [*Commentary*] published a great many articles on X's side. . . . I mean X's ideas are perfectly reasonable sometimes, but he has not changed for about thirty or forty years now and you get into this kind of thing and you don't interest people like me.

Because it signified the breaking up of an old tight package, feelings ran very high on this issue. For example, Dwight Macdonald wrote an open letter to Michael Harrington in the

New York Review of Books chastising him for his support of the union. For many years they did not speak to each other (though they are once more on friendly, if not intimate, terms). Similar splits occurred throughout the intellectual community with one of the most notable being between Jason Epstein and Norman Podhoretz.

The war in Vietnam was also divisive, though in the end the issue became one of militancy rather than "line." The same people who favored the Blacks in Ocean Hill–Brownsville complained about the old socialists on the Left, and unions were again cast in the role of villains.

> There's a funny business about this whole radical movement since the fifties. You have a magazine like *Dissent* and although they raised the banner of socialism in the McCarthy days when nobody was listening, I don't run into any of them on the picket line. I mean, the Vietnam War in the last five years has been very curious. You have complete amateurs like Dr. Spock who have come up from scratch and have gotten more and more radicalized. But then all these professional Marxist revisionists – you know they come out mostly from Trotskyism but they are not Stalinists and are probably men of good will; [but] because of their involvement with this myth of the labor unions, they have not taken part.

To be sure, the *New York Review of Books* seemed strongly pro–New Left only in 1967–1968, and to some members of the Left was never really radical.[7] But whatever the labels, none of these ideological divisions is "merely" a battle over ideas. The controversy extends into friendship and personal association and so deeply affects the character of intellectual networks and circles. The following summarizes our entire analysis.

> Ideally, one should see people regardless of whether you agree or not, but in fact that is very hard. It [disagreement] tends to affect one's social life. The relations I had with the people around Journal X had the character of semisocial relations. That's been broken. . . . In the last few years they were diverging in a direction I did not like and printing people like Y whom I abominate. . . . We [the editors of Journal X and himself] were never friends but we would see each other socially. That's over with. I suppose that if we ran into each other there would be some tension.

How accurate are the intellectual elite about the facts of life in their own circles? A composite picture drawn from their impressions suggests that New York City is still a focal point of intellectual life, although its prominence is fading. Many other centers, mainly in the universities, now exist. Although leading intellectuals continue to talk with each other, many are lone operators who have little contact with other important intellectuals. Even in major centers, intellectual social life is less rich than it had been and many old alliances have broken down because of controversies engendered by the New Left and by the war in Vietnam. Until very recently, it was impossible to know whether these subjective views about the nature of social interaction in large groups of people were anywhere near the mark. Modern mathematics and modern computers, developments anathema to many intellectuals, have made it possible for us to gain a better picture of intellectual life.

Rather than merely accepting an observer's own synthesis of the entire world that he sees about him, we began by asking each person to report only about his relations with the people immediately surrounding him. Among other questions about the prestige and influence of intellectuals, we asked each respondent to tell us whom he had talked with about the social problems or issues of most concern to him, with whom he had discussed social-political issues in general, and with whom he had discussed cultural issues. We also asked whether he felt close to any particular intellectual group and noted if he named any names.*

* The specific questions were:
1. Since your particular interest in [X issue], have you talked with any particular person or persons who have had a special impact on your thinking?
2. In the last six months, with whom have you most often discussed [issue X] in depth?
3. In the last six months, with whom have you most often discussed general social-political issues in depth?
4. In the last six months, with whom have you most often discussed cultural issues in depth?
5. What major groups do you think make up the intellectual community in the United States? Which group do you feel closest to (record names as well as groups)?

Despite the almost limitless possibilities (recall that there were 8,000 or so persons who had written for leading intellectual journals in a five-year period), our respondents actually could think of only 232 different persons in answering these questions. But the spread of names of people intellectuals talk with is much greater than the spread of names given when prestige or influence is in question. Only a dozen intellectuals were named four or more times as interaction partners, with another seven named three times. Despite this apparent spread, a large proportion of the central intellectuals with whom our intellectuals talk are in our sample. (Ninety percent of intellectuals named three times or more, 25 percent of the 46 named twice, and just over 10 percent of those 167 named only once are sample members.)

We cannot give a list of top intellectual conversationalists because that would violate the promise of confidentiality we gave to our respondents. A reader might be able to infer from this list and other information to be presented shortly who actually talked with whom. Nonetheless, there is much that we can say about the characteristics of the stars. First, it should be clear that being active in intellectual circles, that is, "getting around" a good deal, is not the same thing as having intellectual prestige, though of course the two are fairly well related (by a Tau B. of .42). We are hardly violating a confidence if we suggest that someone like the late Edmund Wilson, a top prestige intellectual, was well ensconced on Cape Cod and rarely ventured forth into the intellectual bourse. Other persons, not necessarily so reclusive, may have an impact via print which is greater than their capacity or willingness to enter into personal discussion. Other than personal inclination, or isolated geographic circumstances, politics and age seem to differentiate between those who are less and those who are more prone to circulate among the intellectual elite. The more prestigious radical intellectuals seem to have many connections outside the group of reigning intellectuals and take pride in spending time talking with people outside the intellectual network. In fact, some of the younger intellectuals in our sample reject the notion of an intellectual circle altogether and see themselves more as journalists than as intellectuals. The result of these geographic, personal, political and generational factors is that of the dozen most active participants in intellectual discussion circles, only two are *New York Review of Books* "regulars" and both also write for other journals. One of the dozen is with *Dissent,* four are with *Public Interest,* three with *Partisan Review* and *Commentary,* and one has a rather scattered though prolific list of writings. Surprisingly, only half of the top "circulators" are in New York, the very same distribution as for the intellectual community as a whole.

Before we jump to immediate conclusions about New York as a center or about the "conservative" views of leading intellectual conversationalists, we must look at the actual structure of intellectual social circles. Here is where modern computers and modern mathematics become so important. For it is simply beyond the capabilities of even an extraordinary human being to keep track of and analyze the very complex networks of relations which emerge when even fifty persons, much less two hundred, are being considered. Asking for detailed information on who talks with whom is a method of research known as sociometry and it has been in existence since the 1930's, but the method was not capable of dealing with large numbers of people. So intellectuals cannot be faulted for not having a clear picture of their circles: no one could have had such a picture.

Such pictures are now literally possible, as shown below. This is a map of relations drawn by a computer. Each number represents a given intellectual, so that if intellectual number 77 (at the bottom right of Figure 1) says he talks with intellectual number 131, the computer draws a line between them to show that they are connected.[8] Because memories are fallible, and respondents can only give us reasonable guesses, we do not really care whether number 131 actually remembers that he talked with number 77. We will take the latter's word for it. Of course, people tend to remember having talked with someone important, rather than with someone peripheral in the intellectual community. But this merely leads to the "stars" being clustered together, for the most part, in the center of the map, a most reasonable situation.[9] In some cases, individuals are so closely related to each other compared to other individuals on the map that their numbers appear right on top of each other.

We have "blown up" this part of the map for our own purposes, but closer analysis would not in any case be quite fair. The point is to get some overall picture rather than to search out details. One final instruction on intellectual-circle map reading. The heavy lines show the actual "circles" of intellectuals as they cluster together, but we shall come to that most crucial analysis after we take a look at the whole picture.

How tight are American leading intellectual circles today? One would like to have maps made twenty-five years ago, or maps of England and France, for the sake of comparison. Of course, there are no such maps, and we shall have to work with what we have. In one sense, the network is quite inclusive. There are no circles of intellectuals that are not in some way connected to other circles. Of the people in our sample of intellectuals, 75 percent are shown on the map. Those who are missing are absent either because they say they talk to no one (a most rare occurrence), or they do not talk with other *prominent* intellectuals. There are 121 persons on the map, only 82 of whom are actually in the sample. The others are present because two or more people in the sample "ratted" on them. That is, any person can enter the map if two sample members claim they talk to him. This one "outside" person thus connects two in the sample. In this way we can analyze not only the circles of individuals whom we sampled, but all the important intellectuals connected with them. Since, to begin with, we have in our sample most of the key intellectuals, we can state with considerable assurance that intellectual circles are relatively closed and intellectuals are highly "reachable." That is, a rumor planted at one end of the map would have little difficulty in reaching the other end.

On the other hand, American intellectuals are not very densely related, thus supporting the notion of only loose affiliation. Of all the possible connections between the 121 persons on the map, less than 3 percent actually exist. One reason the total map is not very dense is that intellectuals in the United States are divided, as one can see, into two main overlapping social circles and several smaller almost disconnected ones.[10]

As one can see from the butterfly-like map, there were at the time of our interviews in 1970 at least six circles of leading

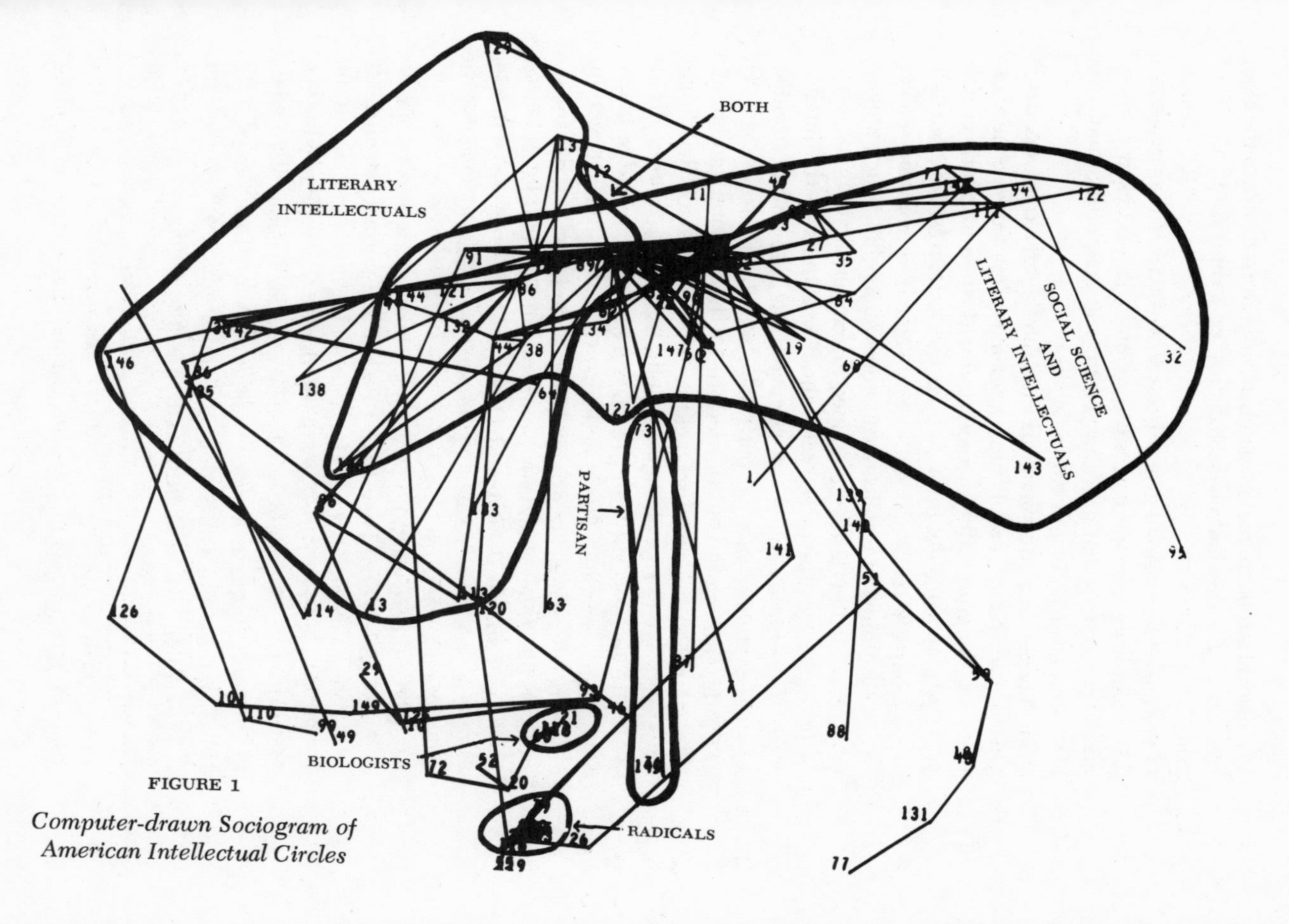

FIGURE 1

Computer-drawn Sociogram of American Intellectual Circles

American intellectuals, of which two were quite separate from the rest. (For detailed characteristics see Tables 11–14.)*

(1) There was a small group[11] concerned with the environment and arms control, consisting of biological scientists, an anthropologist and a leading editor. Because of its small size, this is a tightly knit circle.

(2) There was also a small radical circle, tightly knit, very much not in New York, having a strong West Coast contingent and almost totally unconnected with the rest of the circles.

(3) A noticeable *Partisan Review* circle still existed, but it is relatively small and not clearly separated from the remaining groups of intellectuals, who are, as shown in Figure 1, arranged in three interconnecting circles.

(4) The largest of these, with about one third of the intellectual elite as members, consisted of both social scientists and literary men, 70 percent of whom are professors. (Of the professors, one third were in the humanities and two thirds in the social sciences or some other such field.) The majority of this circle, especially its social scientists, do not live in New York, and there is a strong Cambridge-Boston representation. The members of what we might call the social science–literary circle characteristically write for the *Public Interest* and for *Commentary*, and to a lesser extent, for *Partisan Review* and *Dissent*. At least several write for the *New York Review of Books*, however, and some for the *New Yorker*.

(5) Then there was a literary circle, three quarters of whom live in New York. They tend to write for journals such as the *New Yorker*, *Esquire*, *Harper's*, and *Atlantic*, but most often for the *New York Review of Books*, the *Village Voice*, or the *New Republic*, but hardly ever for *Commentary* or the *Public Interest*. As one might guess, more than three fourths of these are nonacademics. The academics, with one exception, are in English. There is also a good proportion of free-lance writers in this group (and therefore our sample is poorest for this circle).

* Tables 11–14 will be found at the end of Chapter 3.

(6) The last and perhaps the most interesting circle may best be called the center circle, for its members overlap with both the literary and the social science–literary circles. This center circle has 70 percent of its members in New York. Generally older (40 percent are over sixty), and certainly much more prestigious than members of other circles (all but one receiving at least two "votes" on non-Vietnam matters and 60 percent members of the Century Club), the intellectuals in this set also tend to write for a variety of the very top journals.[12]

We are now in a better position to evaluate the impressions intellectuals have of their own circles. The situation is indeed more atomized than one might suspect, lending support to the notion that the fabric of intellectual life had been weakened by the sixties. Consider the matter of the structure of leadership. With the exception of the social science–literary circle, none of the other circles is highly structured in the sense of having a hierarchy of leadership. All the intellectuals chosen six or more times are members of the social science–literary circle and are responsible for the confluence of lines in the center of their circle (see Figure 1). The literary circle is less hierarchical, and in any case all of its leaders are really in the center circle and therefore also members of the social science–literary circle.

There are of course two kinds of intellectual circle leaders: those who are central to but one circle and those chosen by members of two or more circles. The breaking apart of the intellectual elite is shown by the fact that most of the leaders are chosen only by one circle and many of these have such well-known, sharply defined political views that they never had a chance to be accepted by members of other circles. Five leaders do have constituencies in both major circles, however, and there are several more who, though not chosen very often, are well connected with both leading circles. These are of course all part of the center circle. Four of these eight are important editors and another has an important role in an organization which helps writers. The others have in the past occupied key positions in the intellectual community. These leaders are gatekeepers and, though not necessarily personally liked, they cannot be ignored by either of the major groups.

Then there is the issue of geographic structure and the centrality of the journals. The axes are not as clear as some of our informants would have it.

Certainly New York is not everything, though the literary and center circles are still predominantly in New York. But the others are not, and the radicals not at all! Most important, circles are not tightly connected with one or two journals, but with an oligopoly consisting of about ten. There is a locatable *Public Interest–Commentary* group at the corner of the social science–literary circle, farthest from the literary circle (see Figure 1), but its members have sufficient links to those writing for other journals not to make them, at least in 1970, a clearly definable clique.

The situation of the *New York Review of Books* is worthy of additional comment. There is no such thing as a "*New York Review* crowd," though more of the literary circle members write for it than for other journals. Not only do its writers also write for a good many other journals (because they are "stars"), but many of its important writers are not even on our map. This is not because they are low in prestige and hence are not mentioned; they *were* mentioned when it came to prestige or influence. Rather, they are not "clubby" and tend to have a wide geographic dispersion, including, most notably, Great Britain. A fair number of *New York Review* authors who are in our sample of American intellectuals are "isolates," that is, they are intellectuals who are not talking with the people whom other leading intellectuals are talking with. Some of this isolation from the national circle of elite intellectuals is caused by the location of many *New York Review* writers in widely scattered university campuses – a fact which may be related to the comment some intellectuals made that universities are destroying New York as an intellectual center. Then too, some prominent *New York Review* writers are in other circles, especially the radical circle.

How does this picture square with the impression given in the previous chapter that *New York Review*, and especially its editors, are very powerful in intellectual life? The power comes in part precisely because the "stable" of *New York Review* writers is so scattered and so relatively poorly linked. This frees the editors from some of the pressures of group life; they can make their own private decisions without much occasion for advice, second-guessing, or the implicit opinion of a social circle. The popularity of the *New York Review of Books* on many American college campuses may be related to its reflection not only of a single New York–based social circle but of more isolated intellectuals, the very situation in which many who inhabit the oases of American campuses feel themselves to be.

In this respect, the *New York Review* differs from more widely circulated but intellectual-minded journals such as *Saturday Review, Harper's, Atlantic, Esquire* and even *New Republic.* While these also draw mainly upon intellectuals in the literary circle, they tend less toward the academically-based isolate, preferring the nonacademic free-lancer.

Then there is the special case of the radical circle. While many in this circle, in fact most, have written for *New York Review,* many have also written for *New Republic* and *Ramparts,* as well as for other less well-known journals. Their strong opposition to the war in Vietnam ties this group together, rather than geographic or journal propinquity.

Journals are still important, to be sure. They still arbitrate taste, politics, and personality. But at the time of our interviews there was a less clear-cut association between circle membership and journal affiliation than some asserted had once existed among American intellectuals. While this lack of clarity may be the result simply of our more definitive measurement of intellectual circles, we are more inclined to believe it to be a sign of changing times.

The Jewishness of the circles was another theme that recurred in our interviews. The actual data are less than convincing that Jews dominate intellectual circles, though again the question of comparisons is important. All in all, about half the members of circles, and about half of those who are not, are Jews. But two thirds of the radical circle were Jews (the only strong indication of Jewish radicalism in our entire study), three fifths of the important center circle, half the social science–literary circle, and about two fifths of the literary circle were Jews.[13] So while

Jewish interests may be important in intellectual circles, they may not be overriding. (For the characteristics of circles, see Table 11.)

In class origin, the literary circle is the only group that stands out as having better educated fathers, and this is mainly because of its greater number of WASPs. This upper-class bias may continue into the present generation, for two thirds of the group for whom we have data report a family income of more than $45,000 per year. But for most of our sample, their income is an earned one. The center circle shows half its members earning more than $45,000 and much of this money may be from royalties and other perquisites of intellectual stardom. The radicals, appropriately, earn the least, with half saying their family income is under $25,000. The social science–literary circle, perhaps because it includes so many professors, also has the largest proportion in what is for our sample a middle-income group – between $25,000 and $45,000 annual family income.

The impressions intellectuals had of their circles were most concerned with politics, and there are indeed differences in the politics of different circles. (See Table 13.) More than half of the radical and center circle members who answered our questionnaire said they were radical when they were seniors in college, though only one third of the radical circle said they were radical ten years ago. Few center circle members were radical ten years ago, nor are many radical today. Rather, two thirds now check that they are "strong liberals." Showing the effects of the sixties, the radical circle is back to its radicalism, however, with two thirds saying they are radical today. About 30 percent of the social science–literary circle checked that they were radical when in college. Like the center circle, they too moved away from radicalism and have not returned. More than half check "strong liberal" today, while one third are moderate liberals or even conservative. There has been a similar trend away from radicalism among noncircle members, except that half are now moderate to conservative. In contrast to other circles, the literary circle has been radicalized by the events of the last ten years. All of those who returned our questionnaire checked somewhere between "moderate liberal" and "conservative" to indicate where they had been politically in college. No one indicated he had been radical ten years ago, though almost half said they were "strong liberals." In 1970, the majority were either radical or strong liberals.

An intellectual's own estimate of whether he is radical or liberal is of course not necessarily an accurate reflection of where he stands on current issues. These words mean different things to different people. Nevertheless, in 1970 the radicals were consistent with their own self-definition: on race issues they tended to agree that the Panthers were a positive force, that basically we have been a racist nation, and that the Blacks should control their own schools. In contrast, the social science–literary circle disagreed with these ideas, except, curiously, on the matter of Black control of schools, which despite or because of the resolution of the New York teachers' strike, they tended to support.[14] The center circle and the literary circle were between these two groups in their opinions on Black issues.

On style in politics, the radical circle again stood out, agreeing that traditional politics does not make for social change in the United States and denying that there is no justification for violence. (See Table 12.) The other circles were all fairly well agreed that violence is unjustified and that traditional politics still has a chance. But if one looks at cultural style, there is an amazing difference between the social science–literary circle and the others. One hundred percent of the radical circle agreed that rock music is an important contribution to American culture, two thirds of the center circle and the literary circle agreed, but less than half of the social science–literary circle were convinced of the importance of rock.

The division between circles on politics and style is related more to the new issues than to old ones. When the present generation of leading intellectuals were approaching intellectual stardom, the key issues of the day were collective control of the economy, Marxism and socialism. And in 1970, when it came to some of the old issues of government control of private enterprise, the elimination of private profit as the guiding principle of industrial production, and other such items which reflect a "socialist" orientation, old habits still persisted. The radical circle still — by and large — approved of "socialism," half of the members of other circles approved of it, but very few of the literary circle did. That is, the literary circle had been radicalized on recent issues, but retained its conservatism on older ones.

To conclude, there are signs of the times in the picture of intellectual circles just given, but also many aspects of the past. *Intellectual circles reflect most of all the resolution of past controversies, not the lineup on contemporary ones.* The major political event of the sixties was of course the war in Vietnam, and an entire section of this book is devoted to it. To get ahead of our story, Cold War attitudes in the late fifties were related to membership in the various circles: radicals opposed the Cold War, most of the literary and central circles favored "containment" and by the late fifties the social science–literary circle was split down the middle. (See Table 15, Page 152.) Noncircle members held all shades of opinion with the majority for containment. Opinions on Vietnam, however, were less related to circle membership. The radicals were in favor of getting out immediately, but otherwise, circle membership did not predict opinion. The pattern of discussion on the war in Vietnam, indeed on any major contemporary issue, by and large, looks more like a network radiating out from a core of experts than like a structure of circles or cliques. Patterns of discussion on the war were for the most part unrelated to other patterns of discussion to intellectual circles. Former friends found themselves taking different sides. Though most intellectuals finally opposed the war, the degree of fervor they expressed, the timing of their change, and their reasons for their change were quite diverse. The "funny business about this whole radical movement since the Fifties" (above, Page 80) is precisely this lack of close correspondence between intellectual circles and positions on the war. The circles we have drawn demonstrated some of the effects of the sixties, but still showed as much of the fifties and late forties as of the current era. For the curious thing about the sixties is that, despite some claims to the contrary, there was no new lasting synthesis of ideas and values in the intellectual community. Without this synthesis, there can be no fundamental realignment of intellectual circles, only some shifting and loosening of bonds. All their echoes of the past suggest that we had better look more closely at the history of the American intellectual elite in the late forties and in the fifties before we move on to a discussion of Vietnam and the issues of the sixties.

TABLE 11

Characteristics of Intellectual Circle Members and Nonmembers (In Percentages)

CHARACTERISTIC	CIRCLE: RADICAL (N=6)	LITERARY (N=10)	CENTER (N=10)	SOCIAL SCIENCE–LITERARY (N=36)	NONMEMBERS (N = 110)
Age[a]					
Under 50	50	50	0	42	31
50–59	33	10	50	31	32
60 plus	17	40	50	27	37
	(N=9)[b]	(N=17)[b]	(N=14)[b]	(N=42)[b]	
Occupation					
Nonacademic	67	77	57	28	69
Humanities professor	0	23	35	25	18
Social science or other professor	33	0	7	47	14
Residence					
New York	17	77	70	39	54
Elsewhere	83	23	30	61	46
Religion					
Jewish	67	37	64	50	45
Other	33	63	36	50	55

[a] This is based on sample only.
[b] The total includes samples of 110, total list of 172, and others named as circle members but who were not in our total list of 172 and for whom information could be found.

TABLE 12

Various Indicators of the Prestige of Intellectual Circles
(In Percentages)

	RADICAL (N = 6)	LITERARY (N = 10)	CENTER (N = 10)	SOCIAL SCIENCE–LITERARY ONLY (N = 22)	NON-MEMBERS (N = 48)
Mentioned as Influential or as Discussion Partner					
Two or more times on Vietnam	50	30	60	31	19
Two or more times on matters other than Vietnam[a]	50	39	93	50	36
Literary Production in Journals Sampled[b]					
Sixteen or more articles	33	20	40	25	18
Three or more books reviewed	67	30	60	44	32
Membership in Prestigeful Organizations					
National Institute of Arts and Letters	17	0	40	8	6
American Academy of Arts and Sciences	17	10	30	42	17
Century Club	0	40	60	19	21
P.E.N.	33	60	20	19	25
Authors Guild	33	40	20	25	35
Importance in Vietnam Network					
"Member" of inner discussion circle	33	30	90	39	17

[a] Figures for these percentages include all circle members, whether or not they were actually sampled. Other lines represent sample only.
[b] See Appendix for details on journal sampling.

TABLE 13

Political and Religious Stance of Intellectual Circles (In Percentages)

	RADICAL (N = 3–4)	LITERARY (N = 6)	CENTER (N = 6)	SOCIAL SCIENCE–LITERARY ONLY (N = 22)	NON-MEMBERS (N = 27)
Present Politics					
Conservative/moderate	33	40	17	33	50
Strong liberal	0	20	67	57	42
Radical	67	40	17	10	8
Politics 10 Years Ago					
Conservative/moderate	33	60	33	34	55
Strong liberal	33	40	50	48	30
Radical	33	0	17	19	15
Politics When College Senior					
Conservative/moderate	33	100	28	29	44
Strong liberal	0	0	14	43	24
Radical	67	0	57	29	32
Religious Attitude					
Indifferent or against religion	75	83	100	59	50

[a] Numbers in this table reflect the lower rate of return of the questionnaires.

TABLE 14

Political Opinions of Intellectual Circles
(In Percentages)[a]

	RADICAL (N = 4)[b]	LITERARY (N = 6)	CENTER (N = 6)	SOCIAL SCIENCE–LITERARY ONLY (N = 22)	NON-MEMBERS (N = 27)
Blacks					
Panthers are a positive force	75	40	50	23	50
Basically we are a racist nation	75	100	71	36	67
Blacks should control own schools	100	100	86	73	89
New Politics, Culture					
Traditional politics can't get meaningful social change	75	17	29	18	21
There is no justification for violence	25	67	83	77	54
Rock music is important contribution	100	67	67	45	56
"Socialism" Index					
High[c]	75	33	57	43	48

[a] Percentages given are for agreement with statement.
[b] Numbers from questionnaires returned.
[c] Those agreeing to all four of the following statements (abbreviated):
(1) Big corporations should be taken out of private ownership.
(2) Business is overly concerned with profits.
(3) Preservation of the environment requires greater control over private business and local communities.
(4) Labor movement should demand larger role for workers in plant management.

4

Intellectuals and American Foreign Policy during the Cold War

THOMAS J. CONWAY

We cannot understand today's intellectuals or their responses to the Vietnam War without first examining their attitudes toward the Cold War, the most important and pervasive concern of American intellectuals over the last twenty-five years. In our interviews we talked about the Cold War at some length with almost all of our respondents and found that an individual's position on United States Cold War policies significantly shaped his response to the Vietnam War and even his outlook on many domestic issues.

To learn how intellectuals had come to view American foreign policy as they did at the start of the Vietnam War I traced their ideas back to the late 1940's and concluded that their responses to the Cold War fell within three broad categories, each with two versions. I term them: the "Stalwart" approach, which contained both conservative and liberal-leftist versions; the "Critical Support" approach, which had a "realist" version and a democratic-radical version; and the "Recalcitrant" approach, which included an independent-leftist version and a Marxist radical version.*

Despite changes in the positions of many individuals during the period 1947–1962 I found that these six approaches remained a constant, and it is this continuity of approaches, rather than individual changes, that I wish to emphasize. The six approaches took final form in the late 1940's and early 1950's; in order to understand the context in which they were developed we shall first have to look at the years immediately following the end of World War II.

The Early Postwar Years

In the turbulent and chaotic early postwar years America's intellectuals were faced with the simultaneous tasks of defining their attitudes toward Communism and the growing hostility between the United States and the Soviet Union, of responding to America's new world position and her atomic power, and of reevaluating their political philosophies to contend with these new realities. The end of the most widespread and most destructive war in history found the intellectuals in divers states of hope and fear, optimism and dread, energy and exhaustion — and in little agreement on how best to confront these tasks.

The war had given rebirth to various forms of Wilsonian internationalism among the intellectuals and a number of them

* I realize that even so many as six approaches fail to catch every one of the intellectuals' fine-hued reactions to the Cold War, for there were never any unquestioning supporters of the official United States Cold War policies nor were there any pure opponents. I also realize that some intellectuals' positions overlap two approaches, that there are varied differences between the positions of intellectuals whom I have included within an approach, and that some intellectuals move from one approach to another over time. To do full justice to the variety and diversity of intellectual opinion on the Cold War, however, one would have to make an intensive, detailed historical study, something which is beyond the scope of this chapter, but is a task I hope to undertake at a later date. For now, the scheme of six approaches, tentative as it is, can be, I believe, a useful tool in attempting to get a firmer grip on the complex reality that is American intellectual history during the Cold War. Finally, I should make it clear that when I speak of intellectuals in this chapter I am dealing with those politically-minded intellectuals who were concerned with foreign affairs.

actively participated in the wartime movement that pressed for the creation of a peace-keeping organization which would prevent future world-wide catastrophes. These internationalists included advocates of a strong and active UN (for example, James Shotwell and Clark Eichelberger), Atlantic Federalists (Clarence Streit), World Federalists (Ely Culbertson), and finally the World Constitutionalists (Robert Hutchins).[1] Although they disagreed on what type of world organization was best, they were agreed that some such organization was necessary to prevent another catastrophe such as World War II. Their fears of Stalinism were secondary to this higher goal. Indeed, most of them emphasized the necessity for cooperation between the United States and the U.S.S.R. as the basic ingredient for any truly workable world order. They conceded that this cooperation would not be easy to achieve, yet they believed that relations between the two new superpowers could at least be normalized.

The use of the atomic bomb added even greater urgency to their efforts and drew many new recruits to their ranks.[2] The bomb also shocked some atomic scientists into forming a movement for international control of atomic energy, and although primarily concerned with this particular issue, many contributors to their journal, *The Bulletin of the Atomic Scientists*, expressed sympathy with the aims of the advocates of a world peace-keeping organization.[3]

Other intellectuals doubted that an international organization, whatever the form, was the best means to prevent future wars. During the war such figures as Walter Lippmann and Nicholas J. Spykman argued against the hope that any type of world organization could keep the peace.[4] They emphasized, instead, that not a new world organization, but a new balance of power would emerge from the war, a balance based on the realities of power and national interest, and that the only way to ensure a postwar peace of any durability was to ensure that the spheres of interest of the major powers were clearly delineated, thereby minimizing the rivalries which would naturally arise.

Another sector of the intellectual community, the anti-Stalinists – those liberal, radical, and ex-radical intellectuals who, although they differed in many ways, were united in unyielding opposition to Stalin and Soviet Communism – were affected by the war in a different way. They were concerned with preserving world peace, but they saw the rise of the Soviet Union to the position of a major world power as the major threat to that peace; any real cooperation with the Soviet Union after the war, they believed, would be difficult and possibly hopeless. They had continued their criticism of the United States as well as Stalinism during the war, but the war had left them with a new appreciation for the idea of an economy that was not completely controlled by the state, a new appreciation for democracy – for American democracy in particular – and an abhorrence for any form of totalitarianism. The liberal adherents of this view, with such spokesmen as Reinhold Niebuhr and, later, Arthur Schlesinger, Jr., had no one journal of opinion that best expressed their views, but early in the postwar years they could be found in the *New Republic*, the *Nation*, and to a lesser extent in the academic-intellectual journals or magazines such as *Harper's*. Their ideas took on an organizational form with the founding of the Americans for Democratic Action in 1947.[5] The anti-Stalinist leftists (for example, Bertram Wolfe and Sidney Hook) spoke through such journals as the *New Leader*, *Partisan Review*, and *Commentary*.[6] A final group of anti-Stalinists, the ex-radical, nascent neo-conservative intellectuals (such as William Henry Chamberlin, Max Eastman, and Eugene Lyons), used such journals as *Human Events* and the *American Mercury* to convey their views, but they also contributed to the *New Leader* and to popular magazines such as the *Reader's Digest*.[7]

Other intellectuals of various political persuasions had, by the end of the war, developed very strong antiwar views as well as trenchant criticisms of American policies, and a strong anti-Stalinist stance. Some supported the idea of a peace-keeping organization such as the UN (one not dominated by the large powers), but they were very wary of the motives of both the United States and the Soviet Union, doubted that peaceful relations would be established between the two superpowers, and generally blamed both for the breakdown in relations after Yalta. Their sympathies lay with those popular, left-leaning (non-Stalinist) movements which were seeking to establish popular governments in Europe or an end to colonialism. The intellectual journals which supported such points of view included *Politics* (a voice for the anarchism-pacifism of its editor, Dwight Macdonald), the socialist journal, the *Call*, the Trotskyite *New International*, the *Progressive*, and to a lesser degree, the journals of liberal, intellectual Catholicism and Protestantism, the *Commonweal* and the *Christian Century*.[8]

The war had convinced still another group of intellectuals, many of the left-liberal contributors to the major liberal journals of the period, the *Nation* and the *New Republic*, that the hope of a peaceful world lay not only in cooperation between the United States and the U.S.S.R. and a strong UN, but even more important, in American support of those revolutionary forces unleashed by the war which were attacking outmoded forms of capitalism and colonialism. They believed that such revolutionary movements should not be denied support simply because some were communist-dominated, but they had great doubts that any United States support would materialize in the increasingly conservative postwar atmosphere. Even though they had many qualms about its form of government they still had sympathy for the Soviet Union because of its role in defeating Hitler; and they believed that its aims were basically peaceful. They were also still willing to tolerate the existence of domestic communists although they looked upon them with distrust.

Finally, the Stalinist intellectuals, through such journals as *Science and Society* and the *New Masses* (after 1948, *Masses and Mainstream*), gave their support to the Soviet Union and its policies for the postwar world. They feared the reaction of America's leadership to the now powerful Soviet Union and began to turn away from the popular-frontism of the war years to open hostility to the leaders of the United States, and girded themselves for their next assault on capitalist America.[9]

The Cold War Is Declared: 1947

As the United States and the U.S.S.R. failed to find a basis for common action in settling the affairs of Europe and in solving the problem of atomic energy control during 1946, and as crisis replaced crisis, each circle of intellectuals began to press its case with ever stronger feelings of desperation, and the intellectual atmosphere, never calm even during the war years, became more turbulent. The announcement of the Truman Doctrine in March 1947 was a signal for the official shift by the United States government from attempts at cooperation to open hostility toward the Soviet Union. The doctrine elicited wholehearted support from the conservative anti-Stalinists, while the internationalists, balance-of-power advocates, and the liberal and leftist anti-Stalinist intellectuals supported it with many qualifications. Most members of these latter circles were agreed that some form of aid to Greece and Turkey was necessary, but they criticized the doctrine from their own perspectives: it was either unilateral and outside the UN, or its emphasis was too heavily military, or it was too global in its implications, or it would be used to support reactionary governments. It forced the antiwar, anti-Stalinist intellectuals to make very difficult choices between their opposition to war, to colonialism, and to Stalinism, and as a result the first signs of division and fragmentation began to appear in their ranks. Some left-liberal and all Stalinist intellectuals opposed it vehemently, castigating it for its false assumptions about the postwar world, for lending aid to British imperialism, and for needlessly provoking the Soviet Union.

The June announcement of the Marshall Plan, the other major United States Cold War policy of 1947, met with a warmer reception. Because of the plan's economic, nonmilitary character, its humanitarian tone, its less-than-global emphasis, and because of (or despite) its obvious political implications and the lack of any clear alternatives, most intellectuals, no matter how reluctantly, gave it their support. Only a handful of intellectuals wholeheartedly opposed it.

The Turning Points: 1948–1950

The following year, 1948, was a major turning point for those intellectuals who remained opposed to Truman's new policies. The Czech coup (in February), the most startling Communist takeover in eastern Europe, and the beginning of the Berlin blockade (in June) frightened most of those intellectuals who were not already highly critical of Soviet behavior and convinced many who were wavering that Stalin had indeed embarked on a campaign to increase his control in Europe. The intellectual critics of United States policy saw themselves as being forced to choose sides, and many, however reluctantly, increasingly chose to support Truman's policies. The hopes for a world without war and for peaceful and cooperative relations with the Soviet Union were being torn apart by the realities of the postwar power struggle. The year 1948 saw the beginning of the end of many of the bright hopes for world government, control of atomic energy, and a peaceful postwar world.

This change in intellectual temper was reflected in the intellectuals' responses to the presidential election of 1948. Henry Wallace, who laid the major share of the blame for the Cold War at the feet of America's leaders and called for forbearance, patience, and good faith in dealing with the Soviet Union, was supported, finally, by only a handful of intellectuals. No noncommunist intellectual journal endorsed him, not even the *New Republic,* the ostensible editor of which he had been from December, 1946, until the time he made his candidacy known in late December, 1947.[10] This is not to say that the intellectuals adored Truman. Most still had many reservations about him and his policies, but their fears of the Soviet Union and of a Republican victory left them little choice but Truman. Some intellectuals, still unwilling to support Truman, supported Thomas, who, although he was not far removed from the President in his hostility toward the Soviet Union, still pleaded for disarmament and a peaceful solution to the tensions between the two powers.[11] Truman's surprising victory lent a certain legitimacy to his Cold War policies. Most of those intellectuals who had remained sharply opposed to his policies throughout 1948 were disheartened, and their energies were further dissipated as they began to retreat before the joint onslaughts of both official spokesmen and their anti-Soviet intellectual peers.

The formation of NATO, the development of the atomic bomb by the Soviet Union, and the victory of the Communists in China in 1949 pushed the remaining critics further onto the defensive.[12] They were dealt another blow with the North Korean invasion of South Korea in June, 1950, and with Truman's decision to extend containment to Asia by intervening in the Korean War and by extending military aid to the French in Indo-China. All but a handful of pacifist, left-liberal, radical, and Stalinist intellectuals supported his intervention in Korea.[13]

The Korean War was, as was 1948, a major turning point both for America and for America's intellectuals. The war seemed to have confirmed much of what the anti-Stalinist intellectuals had been saying since 1945; for many it was proof enough that Stalin was engaged in a concerted effort with the help of Communist parties throughout the world to extend the power of Russia and Soviet Communism.

The Pattern Is Set: The Early 1950's

Many of the intellectuals had thus made some kind of peace with America and with at least the basic outlines of America's Cold War policies, even as Joseph McCarthy emerged to add salt to the still unhealed wounds of the past decade and to both ride and foster a wave of anti-intellectualism that would last throughout the 1950's. In this atmosphere some influential circles of intellectuals began to display an almost unshakable faith in the good intentions and good deeds of the United States government in foreign affairs. In other circles criticism of United States Cold War policies did continue, but dissenting opinion began to be considered respectable only if it was made within the framework of agreed upon anticommunist fundamentals.

From the choices the intellectuals were faced with in the late 1940's and early 1950's – whether to support or oppose official United States Cold War policies, whether to give them qualified or wholehearted support, what were the best means of opposing communism – three broad approaches toward the Cold War emerged. Various versions of the first two approaches were to dominate intellectual discussion of the Cold War throughout much of the following decade and would help to shape the views of many in the generation of younger intellectuals in the 1950's, while the third would gain intellectual respectability only in the late 1950's and early 1960's.

The Stalwarts

The first approach, which I call the "Stalwart," had many adherents in the early 1950's who, although they held divergent political philosophies and disagreed on the best means of opposing communism, were agreed on certain basic premises. A central premise was the belief that Stalin's Russia was of the same character as Hitler's Germany – a totalitarian state which could not change, except for the worse.[14] Another was that the Soviet Union was committed (because of its ideology and Stalin's lust for power) to the cause of world chaos and revolution and the expansion of its own power either through aggression or subversion. Communism was monolithic. All Communist parties throughout the world, subject as they were to Stalin's will, were united in his aggression. Only the United States had the power to contain this menace. Despite its many imperfections, the United States was the world's last best hope. Indeed, some Stalwarts argued that by simply performing the negative function of containing communism, the United States, at the same time, could and would act as a force for positive good throughout the world.

Premises such as these were accepted, for example, by many members of the American Committee for Cultural Freedom. Its select membership was composed largely of anti-Stalinist liberal and leftist intellectuals, but also contained a scattering of conservatives.[15] Although it was organized as a loose but still united anticommunist front, there were many tensions and the organization gradually faded as a united front as its members came to blows over both McCarthyism and how bellicose a United States Cold War policy they would support.

On McCarthyism, for example, a minority, the conservatives, supported both McCarthy's goals and methods. To them anti-McCarthyism was simply softness on communism and another way of playing into the hands of the sworn enemies of the United States.[16] The majority, on the other hand, were critical of McCarthy as well as the hysteria over communism that produced him. They saw him as a demagogue, with little real interest or knowledge of communism, who exploited popular passions against a hardpressed governing elite. He discredited true anticommunism at home and abroad and hindered honest efforts to root out the communist evil. A number of these liberals and leftists went further and condemned him as even a greater danger to American freedom than communism.[17]

On the question of how best to deal with the aggressive power of Soviet Communism, the Stalwarts also had significant disagreements, despite their accord on basic premises. The more conservative Stalwarts believed that the best the United States and the West could hope for would be only to hold their ground and never to make any gains. It was the United States that was being contained, not the Communist powers. The conflict should be turned into a crusade for democracy and the United States should intervene in the Communist sphere when opportunities presented themselves. To do this, the United States would have to continue to develop new weapons as well as an overall aggressive strategy to fight the Cold War. As for the "Free World," these Stalwarts showed more concern with keeping countries in the "Free World" than with such countries' forms of government, no matter how nondemocratic. The UN, they argued, because it contained Communist members who would only disrupt United States policies, should be avoided and the United States should control its own foreign policy with its own military alliance system. Finally, negotiations with the Communists should be avoided if possible since they were useless; if they had to be engaged in, they should only be used as political weapons for propaganda purposes.[18]

The more liberal and leftist Stalwarts supported the policy of military containment (as a regrettable necessity) especially when the Cold War was centered in Europe, but as the Cold War dragged on, many, although still supporting containment, began to decry the simplistic anticommunism expressed by Eisenhower and Dulles and the increasingly singular reliance of the United States on military solutions to contain communism.

Many, in fact, moved over to the approach I later call "Critical Support."

For the liberal-leftist Stalwarts' purposes, the "Free World" was a useful way to describe those nations that the United States was committed to defend. Although it contained non-democratic states (which should be encouraged to become more democratic), it was agreed that they should be defended, for communism was certainly the greater threat. The UN, they believed, should be supported in its attempts to preserve world order and its economic and relief programs should be encouraged, but it should not be depended upon to solve those questions where the vital interests of the United States or the West were concerned, for the solution to those questions depended ultimately on America's own armed strength. The arms race was the tragic price the United States had to pay for freedom, but economic aid (as a political weapon) could not be neglected. They insisted that the struggle against communism (it was not simply a struggle with another state) was inescapable, but while they favored U.S. intervention where necessary to contain this ideology they were not as bellicose as the conservatives and were generally more averse to an openly aggressive anticommunist crusade. Finally, they agreed that although nothing much could be expected from negotiations, as the communists generally used them only for political purposes, they still might be used to achieve some limited ends and thus should not be completely rejected. Because many had made a certain peace with America and had come to see it as the world's last best hope against the spread of totalitarianism, and since their energies were increasingly channeled into problems of foreign affairs, many of them began to neglect fundamental criticism of domestic America and its growing military power.[19]

The Critical Supporters

The second broad approach, which also had wide influence, can be termed "Critical Support." Under this rubric can be listed such diverse critics as the so-called "realists" (who were generally of a liberal persuasion) and some independent radicals and democratic socialists. The Critical Supporters agreed with the Stalwarts that the Soviet Union and Stalin were dangerous and that some form of containment was necessary, but they had strong reservations about the United States government's interpretation of containment. Within a respectable consensus of anticommunism, they offered some alternative interpretations.

To the advocates of "realism" the discussion of Cold War policies had to center on the concepts of "national interest," "limits of power," and "balance of power." From the perspective of national interest they criticized the leaders of the United States as confused and uncertain over the ends and means of foreign policy, indeed over the very nature of foreign policy. They decried the government's lack of discrimination, its oscillation between isolationism and globalism, and its insistence upon defining foreign policy in sweeping ideological or moral terms. United States policies showed, in these critics' eyes, a lack of a clear or consistent view of what America's role in the world should be and of what were the vital interests that were worth defending. The United States should be alert to defend itself or its allies when its national interest was clearly at stake, but it should not indiscriminately intervene in the affairs of other nations.

The "realists" also emphasized that the power of the United States had its limits, that America was neither the world's policeman nor did it have a monopoly on the world's wisdom. A strong, armed defense was necessary, but they urged that it not be used indiscriminately or wasted on foolish adventures. The United States had to define exactly what it wanted and use its power only in those areas where it would be most effective. By following the Truman Doctrine and the Dulles policies, the United States, they claimed, had engaged itself in an essentially futile and even self-defeating foreign policy – a policy of trying to contain communism militarily all over the globe. They gave their support to United States policy on the Korean War, but they urged that the war be kept strictly limited to that peninsula.

Finally, they employed the concept of the balance of power.

They agreed that the Soviet Union was a totalitarian state, but they argued that its goals of expansion were limited and that the sources of its expansion were not only ideology and Stalin's lust for power but also traditional Russian national goals and the natural desire of a powerful state to expand its influence. The Soviets were not out to conquer the world, but they did want to expand their power. The task of the United States was to teach the Soviets the limits of their power — through containment. The United States had to keep up its armed strength but should not rely solely on military containment, for it was an ideological struggle as well, and political and economic weapons were also essential, and in most cases more effective than military ones. They argued that a modus vivendi should be the ultimate goal of United States policy, in order to shift the competition between the two blocs from the military to the economic and political realm. Although relations could never be cordial between the two blocs, they insisted that they could at least be normalized. To do this, negotiations were required and they could and should be entered into when possible and from a position of strength. They kept in mind, however, that vital interests could never be negotiated, and thus there would always be some tension between the contending camps. The advocates of "realism" were generally European-oriented, and such concepts as the "Atlantic Community" or the "Western Alliance" were more important to them than broad abstractions such as the "Free World," and while they were supporters of the UN, they ultimately put their reliance on the power of the U.S.

Among their major spokesmen were such intellectuals as George Kennan, Walter Lippmann, Hans Morgenthau, and to a lesser extent, Reinhold Niebuhr. Journals such as the *New Republic* (beginning in the early 1950's) and the *Reporter* advocated this approach as did numerous contributors to the journals of the liberal-leftist Stalwarts.[20]

Another version of critical support was offered by independent radicals and some democratic socialists who wrote for numerous journals in the 1950's, but whose position can probably best be found in the journal *Dissent*.[21] Although strong anticommunists who believed that Stalin was responsible for the Cold War and that the Soviets had to be contained, they disavowed the reliance of the United States on exclusively military solutions, its aggressive belligerency and its policy of indiscriminately aiding all anticommunist regimes simply because they were anticommunist. They proposed that the United States, existing as it did in a revolutionary age, had to start to support movements for social change, radical social change as the best bulwark against communist expansion, for right-wing governments encouraged rather than discouraged popular movements to adopt extremist measures, even communism.

As for McCarthyism, both the "realists" and the democratic radicals and socialists opposed it. Many of the "realists" were closer to the liberal-leftist Stalwarts – they criticized the hysteria that the demagogic McCarthy aroused for its interference with true anticommunism at home, his effect on the image of the United States abroad, and for his veto power on United States foreign policy. A good number, however, were also concerned with what McCarthyism was doing to basic rights at home and to the ability of even noncommunists to voice dissenting views on United States foreign policy. The democratic radicals and socialists condemned the hysteria the Cold War engendered and its generally deleterious effects on freedom of speech, on movements for reform and economic welfare, on the ability to profess publicly any kind of radicalism, and on intellectual life itself.

The Recalcitrants

The third broad approach, the "Recalcitrant," slowly grew in influence in the 1950's but only began to become truly respectable with the growing disillusionment over the role of the United States in the Cold War in the late 1950's. The first version of this approach, which had the allegiance of a variety of left liberals, pacifists, independent radicals, and anarchists, argued that even though there were intense ideological differences between the West and the Soviet bloc, the Soviets were not essentially aggressive. Stalin's expansionist objectives were restrained and based on his fear for Russia's security. The responsibility for the Cold War, they maintained, lay with both the United States and the Soviet Union, both of which had divided the spoils after World War II and then fought over the division. In fact, some argued, the bellicose overreaction of the United States to Soviet moves in Europe was an excuse for Stalin to develop a get-tough policy toward the West. This only added to the tensions. These Recalcitrants asserted that nationalism and movements for social change, not communism, were the most important forces in a revolutionary world, but the United States had abandoned these revolutionary forces to Moscow and had taken upon itself the thankless task of trying to sweep back change. Some advocates of this approach held that the United States was indeed engaged in leading a new counterrevolution by its refusal to accept the imperatives of the postwar world and was supporting reactionary regimes (in South Korea, Formosa, Turkey, Greece, Spain – much of Latin America) in the face of the popular leftward movement of the age.

These Recalcitrants criticized the unilateral moves (which they considered outside the UN charter) of the United States, argued for a drastic reduction in the United States reliance on military solutions to Cold War problems, and demanded an end to its belligerent, inflexible, and totally negative approach to world affairs. Although supporting economic aid to Europe and the developing nations, they emphasized that such aid would not be effective unless the United States supported thoroughgoing reform in those nations to which the aid was directed. Negotiations, they pleaded, should not be reluctantly agreed to but should be pursued actively and in good faith. Also, both the Soviet Union and China should be accepted as members of the human community, and the United States should deal firmly, but fairly, with them and avoid bellicosity. A modus vivendi, peaceful coexistence, and peaceful competition should be the goal of United States policy vis-à-vis the Communist powers. But this was not enough – the United States had to start supporting movements for social change throughout the world if it hoped to have any influence during the second half of the twentieth century.

These Recalcitrants also condemned the impact of the Cold War on the United States, the hysteria which led to the repression of civil liberties, the growth of the military and its enlarged voice in foreign policy, and the increased dependency of the economy on arms manufacturers. They were avid anti-McCarthyists, and although they remained distrustful of domestic communists they generally believed them to be of no real danger to America's freedom or security.[22]

This version of the Recalcitrant approach was advocated by many intellectuals in the circles around the *Progressive*, the *Nation*, the *New Republic* (until the early 1950's), and *Liberation* (after 1956), and by I. F. Stone in his *Bi-Weekly*.[23] The persistence of these journals, the *Nation* in particular, in this type of criticism led to loss of income, subscribers, and contributors.[24]

In another version of Recalcitrance a handful of radical Marxist and Stalinist intellectuals, although they were in agreement with some aspects of the first version, placed the entire burden of guilt for the Cold War on the United States. They argued that because of the contradiction of its capitalist economy and the greed of its capitalist class after World War II its leaders found it both necessary and desirable to establish an American empire. They emphasized that the Soviet Union and the other communist states, for which they expressed sympathy, were acting only in defensive reaction to America's belligerency and arrogance. The only way the Cold War could be ended was through a radical (Marxian) change in the economic (and hence, political and social) structure of the United States. Such journals as the *Monthly Review, Science and Society,* and *Masses and Mainstream* gave support to this version.[25]

From the Mid-1950's to the Early 1960's

By 1957 the tensions of the preceding seven years seemed to dissipate slightly. Stalin had died and the Korean War had ended in 1953; the Indo-China War came to an inconclusive end and McCarthy began his decline in 1954. Change was in the air. The Eisenhower-Dulles policies, however, did not respond to such changes and it increasingly appeared that the Cold War had taken on a life of its own. To a growing number of intellectuals the Soviet Union no longer seemed to be the threat that it had been in the late 1940's and early 1950's, while the continuation and global expansion of military containment by Eisenhower and Dulles seemed increasingly at odds with reality. To many critics the policies that were necessary in the late 1940's and early 1950's were now being misapplied under the misguided (but not necessarily evil) leadership of Eisenhower and Dulles, while the more recalcitrant critics tended to see such policies as an outgrowth – a logical extension – of earlier policies. Despite such differences, there was a growing consensus among intellectuals that the United States was becoming increasingly repressive and obsessively concerned with preserving the status quo. Such changes in attitude led to an increase and extension of criticism by the Critical Supporters, a shift of a majority of liberal and leftist stalwarts to the Critical Support approach (a minority of liberals and leftists [e.g., the *New Leader*] continued to advocate the Stalwart approach), and to the growth in influence of the Recalcitrant approach.

In this context, such issues as the development of polycentrism in the Communist world after the death of Stalin, the growing neutralization movement of the emerging nations, and the increasing threat of atomic war by long-range missiles were hotly debated in the late 1950's and early 1960's.

In the face of the rise of polycentrism most conservative and the remaining liberal-leftist Stalwarts continued to stress the aggressive nature of the Communist powers. Although they had to agree, as they moved into the late 1950's, that communism was no longer monolithic, they denied that polycentrism should make much of a difference to American policies. Communist countries differed only in details, the Stalwarts insisted, for these countries were still in accord in desiring to substitute communism for democracy throughout the world. Although the communist leaders (who, it was emphasized, were still Stalinists) employed milder rhetoric and tactics that appeared less aggressive, these men were actually as ruthless as Stalin (witness Hungary in 1956) and, moreover, had world aims that went beyond Stalin's. Although control of all of Europe and Asia was no longer one of the communists' priorities, thanks to containment, it was still their ultimate goal, and the United States had to keep its guard up higher than ever to prevent them from attempting to realize this goal.[26]

The adherents to both versions of Critical Support were warily optimistic about the possibilities of polycentrism. They argued that as the communists became more divided and hesitant, they would look for a detente. This would provide a timely opportunity for negotiations. George Kennan, in late 1957, urged that the major powers "disengage" from Europe — that is, pull back their armies and neutralize at least middle Europe and decrease the military emphasis of their policies as a way of opening larger negotiations over other aspects of the Cold War and avoiding the danger of nuclear war. This proposal was widely debated at the time, but the continuing Berlin Crisis, which erupted again in 1958, tended to dampen further discussion of it.[27]

The Recalcitrants saw the death of Stalin and polycentrism as a chance for the United States to break away from its illusions and the anticommunist stereotypes of the past and begin to come to grips with the many changes in the world since the Cold War began. In 1959, after witnessing the changes that were occurring in the Soviet world, particularly in eastern Europe, C. Wright Mills pleaded for the intellectuals to take a leading role in trying to bring about an end to the Cold War; pleas such as his began to be increasingly influential, in particular among younger intellectuals.[28]

In general the Stalwarts responded to the growth of neutralism and of strong nationalist tendencies among the growing numbers of emerging nations in Asia, Africa, and the Mid-East with dissatisfaction. The conservative Stalwarts insisted that the emerging countries which did not support the United States in this struggle should not be given any form of aid, for neutralism was essentially procommunism. The liberal-left Stalwarts insisted that neutralism was a political evil, but to cut off aid might drive them into the communist camp. They urged that economic aid be used as a political weapon to win these countries over to America's side. The Critical Supporters emphasized that the emerging nations had to be handled carefully. They should be kept as much as possible as allies, but should not be forced to choose sides. The United States was limited in what it could do, and only where its national interest was involved directly should it come into conflict with such powerful nationalism.

The position of the Critical Supporters was that the emerging nations, in Asia in particular, were quite unlike Western nations, and the containment policies that were applied in Europe would not work in Asia. Here the danger was not Soviet military expansion but rather ideology connected with the example of rapid industrialization which the Communist countries provided. The forces one had to contend with were not so much communism as nationalism and poverty, and to do so the United States had to render these countries economic more than military aid, be patient with their nationalist resentment, respect their growing neutralism, and help guide them in a democratic direction.[29]

The Recalcitrants, in particular the independent radicals, saw the growth of nationalism as a force that should be supported by the United States, not condemned by it, for they strongly believed that the emerging nations should be allowed to determine their own futures. To stand against this would be to stand against the powerful force of nationalism and would lead to further United States isolation from the world. It was also considered by some critics to be a positive force that might force the superpowers to lessen tensions and come to their senses about the realities of the world.

Finally, on the issues of atomic war, the arms race, and disarmament[30] the conservative Stalwarts refused to favor efforts to reach an understanding with the Soviets — even in the long run. Instead of talking about disarmament, they argued, the United States should increase its arms and use its technological ability to keep ahead of the Soviet Union. The liberal-leftist Stalwarts believed that a surprise attack by the Soviet Union was unlikely, but increasing levels of violence and small wars that might lead to the threat of big wars were probable. They believed in the need for a strong defense but did not approve of the Eisenhower-Dulles reliance on deterrence. They emphasized that the United States had to be more flexible, and they argued for the development of tactical nuclear weapons and mobile strike forces to be able to fight limited wars but at the same time – it was hoped – avoid widespread atomic destruction.

The Critical Supporters depreciated deterrent strategies, but did not want to abandon deterrence until the United States had something better to replace it. They believed that it was necessary to move toward control of atomic weapons and detente gradually, in small steps, none on a unilateral basis, and with each step conditional on the success of the previous one. Many of the Recalcitrants did not favor military deterrence, even in the short run, and thought the danger of atomic destruction too great for gradualism. They argued that immediate steps, even unilateral atomic initiatives, had to be taken to end the arms race. Only a handful advocated unilateral, total disarmament.

By the beginning of the 1960's, the Critical Support and Recalcitrant approaches dominated intellectual discussion of the Cold War, and Kennedy's faltering attempts to move away from the Eisenhower-Dulles policies met with general approval.[31] After the near nuclear war of the Missile Crisis of 1962, the heated tensions of the Cold War appeared to cool down and the Nuclear Test-Ban Treaty of 1963 seemed to presage the beginning of the end of Cold War hostilities. But the breathing space was brief, for the Vietnam War soon burst onto the intellectuals' collective consciousness and forced them to realize that the Cold War, although it had taken a different form, was still very much alive.

Conclusion

The foregoing analysis was based on the written record of the intellectuals – their articles and books. Where did our own respondents stand during the Cold War? Their reflections on the views of United States Cold War policies held in the 1950's and the early 1960's add subtle confirmation to the written record.

Let us first look at those respondents who were most unreservedly in favor of the Cold War (and who still find it an accurate appellation for the current situation as well). Here are some examples:

I think we should be a shield for areas threatened with takeover due to Communist imperialism. This is the major aggressive force in the world today.

I think we have a major interest in maintaining international stability. We have an interest in preventing international war from becoming chaos. And I think we have an obligation to ourselves, fundamentally, to make sure this doesn't happen – possibly by altering it, and if it does happen, by repelling it.

I don't think the Cold War is over. [The United States] has an unwillingness to face up to its obligations. It wasn't accustomed to a world-wide role and finally lost its nerve. The Soviet Union has followed a consistent path with a clear view of general history, and though its tactics have changed since Stalin's death, nevertheless strategic goals have been pursued with remarkable consistency.

This is the Stalwart approach; 18 percent of our respondents still hold to this point of view.

One-third of our respondents argued that United States policy was essentially correct at least up to the development of the Eisenhower-Dulles policies. They then began to believe that United States policy failed to adjust to the facts of a less monolithic communism and the important role being played by nonaligned states of the "Third World," and shifted to the Critical Support approach. The following respondents would fit into this category:

I don't see how the United States could *not* have been involved in a Cold War with Russia. I don't support those revisionist historians who think the whole Cold War was entirely a mistake. . . . I think people who try to blame the United States for that are crazy or sometimes liars. [He favored the Berlin Airlift and Korea, and supported Kennedy during the Cuban Missile Crisis but opposed the Bay of Pigs invasion.] Now the big question is . . . was it necessary to spend as much money on military equipment and on sending military equipment to other parts of the world where people were supposedly fighting communism? . . . It could be that much of that was wrong.

The Cold War was brought about by the Soviet expansion in Europe. . . . Certainly until Stalin's death in '53, take the period '45–'53, I was in favor, and in retrospect I would still be in favor of, resistance to Soviet expansion in Europe.

There were obvious provocations to which this country responded, though some of the byproducts [domestic hysteria] were unfortunate. There was miscalculation during Dulles's time.

Fifteen percent of our respondents emphasized nonmilitary containment and the "limits of power"; this is the "realist" version of Critical Support. Many of the respondents taking this position mention George Kennan. The following respondents, for example, would belong in this category:

I've never had a simple hard line . . . because I never believed that it [communism] was a monolith. . . . I've never taken to conspiracy theories. . . . You had a much more diffuse and ideological phenomenon which you don't in a sense contain by military means. . . .

Basically you had a situation, particularly in the late forties and early fifties, symbolized to some extent by Czechoslovakia and by the Berlin Blockade, of a pressure against Western Europe. . . . There is never any sense that you win by military power. . . . I think I've always shared the containment policy of Kennan, not in the way, as Kennan himself pointed out, it came to be used, but the thinking behind it: namely, that you create a shield and wait for the internal forces inside the Soviet Union to manifest themselves.

Another version of Critical Support, the democratic-radical version, was stated by this respondent:

On the whole I've been a supporter of the American position taken in the Cold War generally, but I have to go back a very long time to find when my support was anything but severely qualified. . . . As far as economic aid is concerned . . . it goes to [nondemocratic, right-wing] governments, and very often is used to shore up these governments. . . . There are some ways it might be avoided, such as the provisions that [there be] required certain kinds of reforms as conditions for receiving aid.

This approach was supported by 8 percent of our sample.

A fifth group, comprising 11 percent of the sample, most of whom are pacifists, anarchists, left liberals or democratic socialists, for different reasons opposed the policies of both the United States and the U.S.S.R. Here are several examples from this quite diverse set of bedfellows:

I think the Cold War was totally fictitious. . . . It's the people of the world against their general staffs. The general staffs all understand one another. . . . All this is to the advantage of the mass of mankind. So my view is not so much about America and the Cold War as about the people of the world against the power structure of the world.

My position goes back to having been a conscientious objector during the Korean War. Internationally, the United States tended to view communism as a military problem to be dealt with by military responses. . . . This was disastrous. I viewed it as a foreign policy without alternatives to communism.

I never did exculpate the Russians. I never thought for a moment that they were entirely free of responsibility for edging this situation on. But I do feel, and felt then, that the major responsibility was ours. . . . We had a great military and economic power position and it seems to me, in retrospect, as it seemed to me at the time, that we could have afforded to be much more conciliatory than in point of fact we were. . . . I correlate the rise of what I call "witch-hunting" activities in the United States with the intensification of the Cold War.

Finally, those most opposed to the United States Cold War policies held some version of the following position:

I felt that the United States was consciously and deliberately misrepresenting the relations among the Communist powers, the threat of Communist power to us, and was doing so in an effort to construct an American Empire. I considered the Korean War an attempt to encircle China, to maintain puppet states on China's flanks so as to undermine the Communist regime there.

Fifteen percent of our sample claimed they held some version of this Recalcitrant approach.

If we examine the percentages of those respondents advocating any of the last five positions, we find that at least by the early 1960's 26 percent were strongly opposed and 55 percent had many reservations about these policies even before the United States became deeply involved in the war in Vietnam.

What then were the responses of the intellectuals to the Vietnam War? What has been the effect of the war on the intellectuals? Has it led them to question their basic assumptions about American foreign policy? Has there been a massive shift of allegiance from one approach to another? Has it radicalized the intellectuals? How has the war affected their views on the future methods and goals of United States foreign policy? Is a new intellectual consensus on United States Cold War policies developing? To answer these and other questions we will have to let the intellectuals speak for themselves in the next section.

II

The War in Vietnam: The Perspective of the Intellectual

5

How the American Intellectual Elite Decided to Oppose the War in Vietnam

According to the grisly official bookkeeping of the American dead and wounded in the Vietnam War, American involvement began on January 1, 1961, and ended on January 7, 1973. This war, never officially declared, was the most important event of the 1960's, eventually coming to overshadow everything else. Civil rights, poverty, youth rebellion, the sexual revolution and the general vast changes in American cultural standards were all affected by the war and eventually dominated by it. It is impossible to understand American intellectuals without understanding their relationship to the war. American involvement in the war came about partly as a result of Cold War doctrines formulated by the American intellectual elite in the late forties and early fifties. The eventual truce of America's longest war was negotiated by a leading American intellectual. Somewhere in between these events fierce opposition to the war developed among American intellectuals long before the rest of the American public had even begun to think about the war, for by the spring of 1965, eight years before the war came to an end, the majority of the intellectuals we interviewed were opposed to the war. At the time our study was conducted, in 1970, 90 percent of the leading American intellectuals said that it had been a mistake to send troops to Vietnam. Only 56 percent of the American public answering an identical Gallup Poll question at the same time thought so. Further, only 16 percent of the leading intellectuals approved of the way Nixon was "handling the situation in Vietnam," whereas at the very time of these interviews and with an identical question, 54 percent of the public polled by Gallup approved of Nixon. Even when compared to the academic community, elite intellectuals were more likely to be strongly opposed to the war. The sample of elite intellectuals was presented with a question on Vietnam identical to that given American professors (see Chapter 1) one year before our interview. When the professors were made comparable to our set of intellectual elite in terms of their ages and religions, we found 44 percent of elite intellectuals who were social scientists advocating immediate withdrawal from Vietnam, as compared with only 28 percent of social scientists among the professorial elite. Similarly, among those in the humanities, 50 percent of elite intellectuals "wanted out" now, as compared with only 33 percent of the same group in the professorial elite. As we shall see, these differences cannot be accounted for by the difference of one year between the times when the question was presented, since most of the intellectual elite had held their positions since at least 1965.

These are the basic facts. What they mean or portend is another matter. Did the Vietnam War actually radicalize the American intellectual elite? Is the picture of intellectuals as a group out of joint with the rest of America merely a reflection of the constant alienation of intellectuals, or did Vietnam reverse the patriotism of the late forties and early fifties? Our view is that the war merely served as a catalyst in moving intellectuals to the left. For as Conway showed in the previous chapter, the movement with respect to foreign policy had begun earlier, in the late fifties, especially after the death of Stalin. The Cold War–anti-Communism package of the late forties and early fifties was already breaking up by the time the war in Southeast

Asia intruded itself upon the American consciousness. In some ways, the Kennedy response was therefore an anachronism. A substantial proportion of the American intellectual elite was ready to be convinced that we had no place in Southeast Asia, that the domino theory was unreasonable, and that it was neither practical nor desirable for the United States to impose its culture or style of government upon the rest of the world.

This is not to say that the change came automatically. The actual dynamics of the change – how it came about, who championed it, who responded, and the reasons given for the change, had a profound effect on American intellectual life, an effect that is still very much with us. In the long run, far from speeding the radicalization or the reradicalization of the American intellectuals, the war destroyed for at least one generation any chance that the maturity and thoughtfulness of the Old Left or the former Old Left would be combined with the dynamism and enthusiasms of the newly radicalized youth to produce a movement for profound change in the United States. We shall see in Section III that the Civil Rights movement, the Poverty movement, the Student movement and the New Left generally were stopped cold by the war, though on the surface of it the very existence of the New Left seemed to be a response to the war. For the way the American intellectual elite reacted to the war effectively barred most of them from assuming leadership of the young, and the fact is that without this leadership the young eventually foundered, although whether in the long run any of this would have made a difference in policy is questionable. These are heavy charges, especially in view of the strong negative reaction to the war on the part of the intellectual elite, and can be understood only through a careful examination of the relation between intellectuals and the war.

The analysis begins with an understanding of the variations in the "line" intellectuals took on the war. Then we shall look into the sequence of events and influences upon intellectuals which resulted in the current state of opinion. We shall note that changes in opinion on the war came mainly as a result of the input of new information and new feelings from *nonleading* intellectuals who were specialists on the subject, and from college students. We shall analyze the influence of intellectual circles and shall see that, despite their apparent avant-garde position, the elite intellectuals were followers, not leaders, at least within their own circles. Whatever changes came about as a result, and whatever their increased militancy, the intellectual elite were not radicalized in their views of foreign policy or the general domestic state of the United States. After this general picture of influences we shall have a close look at the thinking and reasoning of intellectuals on the war – the aspect of intellectual life that counts the most. Very few took a radical position in explaining how the United States entered the war. In these pre–Pentagon Papers interviews, most intellectual leaders did not see the war as having resulted from the basic economic structure or ideological stance of the United States. Rather, they embraced what were essentially liberal-realist explanations. And their basic objections to the war itself were, even in 1970, overwhelmingly pragmatic, rather than ideological or moral. Matching this, we shall see that intellectuals who influenced other intellectuals were generally *not* radical in their views of the war, though they themselves thought that the case was otherwise. Finally, as might be expected from all this, their views of future foreign policy are at best murky and hardly revolutionary.

What Policies Elite American Intellectuals Advocated for Vietnam

President Nixon, during his first campaign for President in 1968, said that he wanted to end the war in Vietnam and finally managed to do so at the beginning of his second term in office. But Nixon and the intellectuals did not agree on methods or procedures for ending the war. In the following pages we attempt to pin intellectuals down to details on their plans for ending the war or dealing with it.

To avoid almost endless discussion and to insure some comparability among respondents we presented intellectual leaders with four fixed positions upon which they were asked to comment. The positions were those which had been presented one year earlier to American professors by Ladd and Lipset (see above, Page 15):

1. The United States should withdraw from Vietnam immediately.
2. The United States should reduce its involvement, and encourage the emergence of a coalition government in South Vietnam.
3. The United States should try to reduce its involvement, while being sure to prevent a Communist takeover in South Vietnam.
4. The United States should commit whatever forces are necessary to defeat the Communists.

In their response to this checklist, American intellectual leaders in 1970 showed themselves overwhelmingly opposed to administration policies. Only one of the 110 intellectuals interviewed took the last position – defeat the Communists. In contrast, more than 40 percent wanted to get out now – position one. A few more, 45 percent, qualified their position somewhat by taking number two – reduce involvement while encouraging a coalition government. Then there were 13 intellectuals who took the third position, which seems closest to that of the administration: reduce involvement while being sure to prevent a Communist takeover. Not only were these a strikingly small minority among the intellectual elite, but, compared with the rest, they tended to have little or no prestige among other intellectuals.

Other than prestige, there were few basic characteristics of intellectuals which predicted their Vietnam opinion. Age, as we have seen, is not important in this elite group, nor is religion, though both were important factors among the elite professoriate studied by Ladd and Lipset. The effect of college students upon their elders was apparent in the distribution of opinion according to occupation: professors are more likely than nonacademics among the elite intellectuals to favor immediate withdrawal. Ideology, not social background, was the clearest predictor of opinion on the war: two thirds of the small radical clique "wanted out" now, another third favored a coalition government, and not one supported the third or fourth positions; the largest support for these positions, almost 20 percent, came from the more conservative social science and literary circle. These circles, as best we can determine, really predate 1965, and so the word "predict" is intentional. The circles formed in response to discussion about the war will be discussed later on.

By far the best predictor of Vietnam position in 1970 was Cold War position in the late fifties, a fact with profound implications which will become more and more clear as our story unfolds. For the moment, the facts are that 70 percent of those who opposed the Cold War in the late fifties wanted to get out of Vietnam immediately in 1970; 60 percent of those who were Cold War hawks in the fifties wanted in 1970 to prevent a Communist takeover of Vietnam. (A Tau Beta of .43 significant at the .001 level shows that the intermediate positions were proportionately distributed.)

Anyone who has ever gone about with a petition trying to get an intellectual to commit himself to a statement written by someone else knows that a mere count of responses to a fixed position conceals almost as much as it reveals. Most intellectuals we interviewed qualified their responses in one way or another and the following serves as a corrective to simple statistics.

Some intellectuals who took position one did so unreservedly, as did this radical journalist:

> The United States should withdraw from Vietnam immediately—yes, number 1.
>
> [Question]: How could we accomplish withdrawal?
>
> [Answer]: Pull the ships up and load the troops aboard.
>
> [Question]: Any other elaborations?
>
> [Answer]: I think that if anyone within the administration actually wanted to do this [they] could figure out the politics. That's a very technical question. There has to be [first] a basic decision.

Others, while fully endorsing the position, wanted to make clear that their wanting to get out of Vietnam right away did not mean that they were especially favorable to North Vietnam, as the following pacifist explains:

> Many people who have my position, as described so far for immediate withdrawal, are people who have more sympathy than I do

with the other side. That is to say, they're people who think that not merely is the war a total piece of idiocy on the part of the United States . . . but that somehow the issue involves the side which is right – the North Vietnamese, against the side which is wrong – us. I myself have no pure illusions about the justice of the side of the North Vietnamese.

Finally, some who did check number one seriously considered number two but rejected it on fairly technical grounds in wording.

Of course, 1 and 2 overlap. You cannot get out immediately. I mean, it's not like you can get out of this apartment. . . . It takes a year to get out. . . . These questions are not very well . . .

[Question]: Could you restate the way you would like to see your position?

[Answer]: I am in favor of civilian government, any civilized government in Saigon which will make peace with the other side and invite us to withdraw.

A great many respondents checked in between one and two, but we classified them as number two in order to make number one more nearly an unequivocal position. Many of these in-betweeners talked about setting a date and then withdrawing, a position which had become prominent after the positions we used had been written. Most combined the idea of a date, however, with that of a coalition government. For example,

I am opposed to the first one because I think that would be catastrophic in the sense that we should be committing perhaps a million South Vietnamese to the vengeance of the North Vietnamese. . . . But I would like to see a specified scheduled reduced involvement in Vietnam and the emergence of a coalition government.

Others checking number two emphasized the idea of negotiation rather than mere withdrawal.

Well, none of them represent my position which is that the United States should negotiate a settlement and withdraw. . . . Well, I guess it is closest to Number Two. . . . Obviously you could never have a negotiation as long as you stick to the present [Saigon] government.

Finally, some seemed quite despairing of any course of action.

> All right, I'll say Number One with . . . qualifications. I don't think it can happen, so it's not a position. It's like saying my position is that everybody in America should have a million dollars tomorrow.

In the end, this respondent checked between one and two.

Even respondents who checked the position which favored preventing a Communist takeover were not uniformly "hawks" by any manner or means, nor were they supporters of the Nixon line. One political analyst who checked position three actually would "Cut our losses and get out, because I think it is an unmanageable situation." A social scientist explained:

> The one I checked was to "try to reduce involvement and to insure against a Communist takeover in the South." I don't like the formulation because first of all I don't know how you can be sure in terms of reduced involvement. The only way you can be sure to prevent a Communist takeover would be to keep the troops in. . . . The position which I favor, which I don't think the administration has been pressing for, is an attempt to get a cease-fire.

Others checking number three were fairly cynical about the possibilities of preventing a Communist takeover. An expert on foreign affairs commented,

> I am probably closest to Number Three but with the qualification that we should try to avoid a Communist takeover while we are *present* in Vietnam. Once we are gone from Vietnam then I don't think it makes much difference.

Naturally, some of those checking number three took a more conventional government-oriented point of view.

In sum, the three basic positions on Vietnam taken by American intellectuals with various emendations and qualifications are: get out now, get out a little more slowly while worrying about what will replace us, and mostly get out while trying to shore up, however temporarily, the status quo in the south. Even among those taking the last position, clear supporters of current government policies were few and far between. The fact that almost all leading American intellectuals opposed the war in 1970 is incontrovertible, even if their positions varied and their policies for ending the war were a bit cloudy.

When Intellectuals Changed Their Minds

Although the overwhelming number of leading American intellectuals finally opposed the war in Vietnam, this was not always the case. In fact, when the basic fate of South Vietnam was being settled in the late 1950's and when the Diem regime was being set up, only 11 percent of the intellectuals were at all concerned about Vietnam and most of these were merely uneasy about American policy rather than opposed to it. By 1962 a total of 38 percent (including the 11 percent from the previous period) were worried or concerned, though again not necessarily clearly opposed to United States policy. Key events which influenced our respondents, by their account, were the Taylor-Rostow report and the subsequent dispatch by Kennedy of American military advisers to South Vietnam. The Buddhist demonstrations against Diem and his eventual overthrow and assassination affected many in our sample, as did the Tonkin Gulf affair of the next year, so that by the end of 1964 the majority, or 62 percent, had became concerned. The Pleiku incident and subsequent bombing of North Vietnam and the escalation of American troop strength during 1965 captured the attention of almost all the rest, so that by the end of 1965, 93 percent of our sample had become concerned about the war in Southeast Asia.

Opposition to the war grew over time. One quarter of the leading American intellectuals in our sample said they had always been opposed to the war and were opposed at least before 1962. Typical is the art critic who said:

> I used to read the *Nation* . . . and for many years they ran editorials and articles anticipating the increased involvement in Vietnam

of the United States and its inability to get out, which they predicted . . . it was about 1960 or something like that, maybe a little later. . . . When I read about it I naturally seemed to agree. I felt in agreement that this was no place for America to be. In other words we were there because of this fixation on Communism.

About one fourth (23 percent, to be exact) never supported American policies, but became actively alert in their opposition to the war only somewhat later — between 1963 and 1965. Many of these respondents found it difficult to pinpoint the time when they took an active stand in opposition. "It isn't as though you had a sudden conversion," one of them pointed out. "I didn't pay much attention to it when they first sent those advisers over that Kennedy sent . . . [but] as soon as the Johnson business started to escalate I from then on felt very strongly [opposed]."

Those who had always been opposed were joined sometime before 1961 by a small group (6 percent of the total) who had once favored American participation in what were then called "counterinsurgency" tactics but who had later changed their minds. Typical of these early switchers was this intellectual who at first supported the Kennedy decision to send advisers, a decision he now thinks to have been a mistake. He supported Kennedy, he said, "because at that time I gave generalized support to the American effort in the sense of containment of Communism, and partly out of support of Kennedy. I had much less critical judgement at that time, much less critical interest and much more generalized support." Even before major escalation, he developed "more willingness to listen to arguments against American policy."

By the end of 1965, these early switchers were joined by about another fourth (22 percent) of the leading intellectuals in our sample who had switched from a position supporting American activities in Vietnam to one which opposed them. As an economist explained:

I was originally guardedly in favor of the Vietnam thing. I really believed that there was an analogy between North Vietnam and South Vietnam and North Korea and South Korea. . . . I felt that the Russians were, so to speak, a very dangerous bunch. . . . With

Vietnam, I thought, yes, this is a case of, more or less, clear-cut aggression by the North Vietnamese against the South Vietnamese, that we were quite justified in going in, and that it would be a relatively minor operation. I then had not read much about it. . . . I had sort of swallowed the Kennedy line. So I've become a dove only the past five years.

Most of the middle switchers, as we shall call them, changed their minds "around 1965 [when] the bombing began. I thought," said one, "that was just disastrous." As a result, by the end of 1965, three quarters of the leading American intellectuals were opposed to American policy. If my figures are anywhere near correct, then Eric Goldman's view that "the intellectuals" were at this time *split* on the Vietnam issue is entirely wrong.[1] The figures he cites were based on letters or on polls of some university faculties. And faculties as a whole, even leading members of university faculties, are quite different, as we have shown, from the leading intellectuals Goldman and I have in mind.

Finally, the last group of late switchers (only 13 percent) had, by 1967, joined those in the opposition, leaving only 12 percent who at the time of our interviews in 1970 still felt that the United States should actively work toward preventing a communist takeover in South Vietnam. A professor who had served the government was typical of the late switchers:

I was not as sure as many of my professorial colleagues that the whole thing was a terrible mistake. I was a little more impressed by the fact that it was a complicated situation and it was pretty difficult to see what you would do.

But by 1967 he had begun to feel that "the war policy had gone beyond the prudent limits of the policy a democratic government can pursue without the consensus that is needed by that kind of policy." He concedes that a speech that he gave at a mass meeting on campus in 1967 in which he first enunciated this position made him "one of the most conservative speakers at this gathering."

Even the few remaining supporters of American policies in

Vietnam had reservations about the way matters were being handled in Southeast Asia.

> In view of the split that has now taken place in the United States because of the halfhearted way we fought the war and our inability to win it on those terms, I think Nixon's policy of withdrawal and a building-up of South Vietnamese forces to a point where they can defend themselves is the only feasible one. . . . Our initial mistake was that we never should have gone into the war unless we were prepared to win it in military terms, which in my opinion we could have easily done. The whole idea of the U.S. fighting a land war in Asia with American troops is not feasible. We're outnumbered 10 to 1.

To sum up: we are dealing with six basic responses to information and influence about the war in Southeast Asia: a group opposed to American policies on the war from its very start in 1960 or even before there was any significant American presence there; a group which had always been opposed to it but became alerted only later; a small group which had first favored the war and then very early began to oppose it; another group of middle switchers; a group of later changers, and finally a small group which never wavered in their support of the war.

How do we explain why some intellectuals fell into one group rather than another? Once more, social background is relatively unimportant. It is, however, more related to the *process* of change than to present opinions. Age and religion do account for something. Not surprisingly, 6 of the 15 intellectuals in our sample who were forty or younger were early opponents of the war. On the other hand, as we have noted before, we do have an older contingent of radicals so that one third of the 27 sixty-one years of age and older were early opponents of the war. It is the middle-aged group which was slow: less than 20 percent of the 62 persons forty-one through sixty years of age were early opponents.

One of the few pieces of evidence in our study for any degree of greater radicalism of Jewish intellectuals and the corresponding conservatism of Catholic intellectuals – a relationship typical of American professors – is the timing of opposition to the war. If we combine all who were always opposed to the war with those who switched into opposition before 1962, we find 60 percent of the Jews, 50 percent of the Protestants, and only a bit more than 20 percent of the Catholics as early or constant opponents. But these figures are not especially impressive since the difference between Jews and Protestants is not very large, and there are only nine in our entire sample who admitted that they grew up as Catholics.

Ambiance as reflected in background does make for a bit of a difference. Thus, New Yorkers who are more affected by the social climate and by journal reading tend more often to be middle switchers, for these are factors which we shall see are keys to being a middle switcher. Occupation, too, is sensitive to ambiance. For example, almost all who always supported the United States government position are either in political science or public law. Most have had some official connection with the government. And all four in our sample who were at the time of the interview either advisers to government or foundation officials or in some such position were late switchers. This may reflect the "it's-never-too-late-to-be-with-it" ambiance often characteristic of government and foundation officials. Then the humanities professors also tended to act according to stereotype. Half were middle switchers, being neither the first to leap nor the last to join.

But once more ideology was the crucial determinate – ideology that had been established *before* the early sixties. All of the radical circle, half of the literary circle, but only above one third of the central or social science–literary circles were always opposed to government policies. Cold War positions held in the late fifties were the most important determinate of switching on the war in Vietnam. None of those who had always supported United States policy in Southeast Asia held any but the first two positions on the Cold War. That is, they favored the Cold War. (See Table 16.) And no one who in the 1950's was against the Cold War was ever anything but opposed to American policy in Southeast Asia. On the other hand, in between these extremes lie those who changed their minds about the war, and these persons suffered some of the most tortured reappraisals. The largest group of 1950 Cold Warriors were "middle switchers," that is, people who changed their minds about Vietnam around 1964 or 1965. To say, therefore, that past attitudes toward the Cold War tell the whole story is to deny the real inner conflict and policy change which the Vietnam War initiated. Whether this conflict led to radical change, however, is another question.

The agony expressed by some who switched fairly late and who had been clearly and totally identified with the cultural Cold War is indeed great, for it demonstrates how historical circumstances can force intellectuals with clear ideologies to alter their conclusions.

> I've written a whole article against this [position] but still it is closest. I'm writing Number One, that the U.S. should withdraw from Vietnam immediately, and I'm writing that that's the closest to my position. But I would qualify it that I'm not in agreement with the vast majority of people who take this position. Because they take this position like that, and they close their minds and consciences to the whole problem of having said "Withdraw," whereas I think this is one of the most tragic decisions which I have ever urged on anybody. . . . We've got ourselves, by a series of mistakes, into a situation where, tragically, we have to make this decision to withdraw immediately. . . . I think we are turning over the whole of Southeast Asia to Communism. I'm a firm believer in the Domino Theory. I would be very glad to admit that we made a mistake, we're defeated, and get out. But I think the easy moral judgment that's made in the country indicates that most people who take that so-called liberal position – I don't like to concede the word "liberalism" to them, I keep it for myself – but I think that people who take this ritualistic position are just not political. . . . I'm sure that if we don't get out soon, and maybe if we do get out, we're going to be inviting fascism in this country. It's one of the chief reasons for my thinking that we have to get out fast.

The same respondent reflects on the Cold War:

> I have never been one of the people who has thought the Cold War was a horror; I thought it was a fact. . . . You're always in a state of Cold War with any nondemocratic country whether it's Fascist or Communist. I think that to talk about the Cold War as an act of provocation or an act of insufficient sufferance of another way of life is very stupid. . . . It's very good for either side not to have the other gain strength. . . . So I thought at the beginning of the Vietnam war that it would be very good if we don't let Communism gain

any more member nations. And so I was in favor of our engagement in Vietnam at the beginning. I learned from the Vietnam war something that is very hard for me to accept, and that is that there are some moments when you cannot stop the other side. . . . It is not in our control to keep the Third World democratic. . . . We may suffer terribly from this. We may be destroyed by this. That may be the historical evolution that we have to face. . . . I hope not.

The Media and the Circle of Influentials

Two thirds of the American intellectual elite changed their minds on Vietnam during the 1960's, according to the account we have just given. Even those who remained totally steadfast in their beliefs – those who from the very start opposed American policies and those who always supported them – have changed some of the ways they have approached the issue. At the very least they have spent much time and effort in maintaining their positions. Though Cold War ideas were already changing in the late fifties, the sixties witnessed a host of new facts and new ideas about America's role in the world and especially about Vietnam – a place essentially unknown to 75 percent of our sample (and to 99 percent of the general public) before the early sixties.

We shall show that the facts and the ideas about Vietnam came from a central core of intellectuals, many of whom are in our sample. These people were beginning to have firsthand experiences with the war and reported these experiences in the liberal and the left weeklies and monthly journals. Although the printed word was chiefly responsible for the opinions on Vietnam of only 35 percent of our sample, these reports were especially influential among the early opponents of the war. This avant-garde was soon followed by the Establishment press of the American intellectual, the *New York Times,* whose news reports from Vietnam affected almost every person we interviewed. Intellectuals we classified as having always been opposed to the war but whose opposition was activated later, between 1963 and 1965, were especially impressed by the *Times.* The rest of the intellectual community were swept along by the course of events, but were especially influenced by those in the intellectual community itself who had already been convinced and were hard at work spreading the gospel. Teach-ins on the college campuses in 1965, a new wave of articles in leading nonradical journals such as the *New Yorker* and the *New York Times Magazine*, and the resulting atmosphere in the intellectual community in 1966 brought along almost all of the rest of the Cold War holdouts.

Much will be made of the fact that the central core of influentials on Vietnam were really not radical in their views, despite the contrary impression of the intellectual community itself. In one sense, however, there was no such thing as an opinion leader or even a group of opinion leaders on Vietnam. Rather, the war was widely discussed within the intellectual community; everyone in the entire circle of discussion was in fact an opinion leader who helped to reinforce the opinions of all his fellow intellectuals. One respondent correctly noted, "The intellectual community agrees with itself on Vietnam, so it's hard to say who has any influence." This is, of course, a slight exaggeration, since during the crucial early period – 1960 to 1963 – the word was carried mainly by Vietnam experts such as Bernard Fall, rather than by intellectuals who carried more general prestige in the community.

The Media

We shall begin our story of influence at the beginning – with the media, but it must not be forgotten that evaluating the quality of a journal article or news story is the essence of being an intellectual. Our respondents were not influenced by anonymous sources: they were not impressed by an article in the *New Yorker*, but by "Dick" Rovere, someone they had known for a long time and respected. It was not *Commentary*, but "Ted" Draper, not *New York Review* but Noam Chomsky, who carried weight. And if it was someone new like Frances FitzGerald, whom the general public had certainly not heard of in 1970 when our interviews were conducted, then at least one respondent talked at length about her and her theories of the importance of the *I Ching* in understanding Vietnam. David Halberstam was unknown to most of the intellectual community when he began his reporting in 1962. When he returned to the United States he worked with Willie Morris and was at the core of an important circle of New York intellectuals, many of whom had by 1970 talked with him about Vietnam more than once.

Let us plunge into a typical interview with a leading novelist.

I read a good many books, and articles in the *New York Times* and elsewhere. Writings of people like Fall and others. I also read from the opposite side, people like Mary McCarthy or Susan Sontag or Chomsky, which really puts my hackles up. But as I say, it is very difficult for me to straddle these worlds – to reach from the world of imagination which is largely my own concern to this other world of public affairs. I don't claim to do it well. However, when I do look into these matters I look into them thoroughly and try to base my opinion upon reading and study.

[Question]: Which of these writings [Respondent shown a list] . . . Did they make you emphatic about the issue of Vietnam, or was it just one in a number of other things?

Well, I suppose the one book that influenced me the most was the one by Draper which seemed to me to make good sense. I read also a good deal of the foreign press; I read English and French newspapers quite often. I often read *Le Monde* and the *London Times.*

[Question]: What intellectuals do you think have had the most influence on the discussion of the Vietnam issue within the intellectual community?

Well, the people that I get to talk to and who seem to know something about it and from whom I get something are certain colleagues in political science or social science around here like [names several prominent people in his university] and people like Raymond Aron in France or Dick Rovere, who seems to me a sensible fellow.

The preceding quotation is a shortened version of one of our shorter interviews. We asked each person in our sample in considerable detail about how he came to make up his mind about Vietnam. In our interviews we probed for their first thoughts, the series of influences upon them, their second thoughts, and how they might sum up the influences upon them. We asked them to name the names of people they had talked with and people who might have indirectly influenced them. We prodded their memories, after getting first impressions, with a list of major events in the history of American involvement in Vietnam, a list of major books on Vietnam, and a list of major journals. All of this probing yielded a great deal of information about how they became interested in Vietnam and how they formed their opinions about the war. Some of the interviews ran over fifteen pages of single-spaced text for this part of the interview alone.

Each person was asked about the various media that might have influenced him — newspapers, journals, books and television. We also asked about more direct influences such as personal experiences. We asked each person to sum up, himself, the influences upon him. When this was not clear or where the interview was somewhat abbreviated (in 16 cases) we ourselves attempted to make this assessment. In the following analysis, each respondent is classified according to the one single most important influence on the way he first came to have his current position on the war in Vietnam.[2]

Let us begin with the mass media. The newspapers, or rather a particular newspaper, was of major importance. The *New York Times* was read by literally 98 percent of our sample, and its importance with respect to the Vietnam issue cannot be overstressed. The editorial pages were not as important as the news itself. The reporting was highly influential, especially the reporting of David Halberstam, who was second only to Bernard Fall in the number of times he was mentioned as an influence on a respondent's own opinion. In fact, the difference between the formal editorials and the actual reporting of the news in 1962 and 1963 was striking enough to have been noted by several respondents. A *Times* man himself said:

> Because the editorial board does feel strongly that our basic policy in Vietnam is wrong . . . it isn't any part of our thought that the view on the editorial page ought to permeate the news columns. . . . It is quite true that for a long period the news editors of the *Times*, as a matter of personal conviction, were very much in disagreement. The fact that they happened to disagree with the editorial board

didn't affect our news columns, either. But our correspondents, as you know . . . many of their news stories, I think, very objectively did provide the raw material on which many of our readers came to the conclusion that the war was a horrible war.

The *New York Times* was influential not only among those who casually follow current events in the newspaper. A leading critic who had spoken with Chomsky, Schoenbrun, and a great deal with Rovere, and who read Draper, Zinn, Mailer, Galbraith, Schlesinger and Shaplen on Vietnam, along with a good many other books, said:

> I was never impressed by what any American has written about Vietnam except people who have been there, like Bernard Fall. And, of course, I've been mostly influenced by the *New York Times* about the Vietnam war.

And a writer who is himself one of the most often cited influentials on the war in Vietnam said:

> I read most of these books [on a checklist of 44 books we gave to each respondent] but I don't know that they made me concerned. They taught me a good deal. They confirmed some of my worst fears. I think that would be a fair way of putting it – maybe the *New York Times* was most important in terms of more information.

To some extent, then, Agnew is right. The *New York Times* had a very important role in aligning intellectuals against the war in Vietnam. Not only was the *Times* important as a general source of information for almost everyone in our sample, but we judge that almost 20 percent of our sample made up their minds on the war mainly because of information they received from the *Times.*

While almost all of our respondents had read some books about Vietnam, it is apparent that the books were not as important as the *Times* or as journal articles. In fact, two thirds of the often-cited books were compilations of materials that first appeared in journals or newspapers.

A professor of political science remarked:

I don't think books have been important for me, I think books I have read have not formed any new attitudes; they have just reinforced certain ones – like Mary McCarthy's book on Vietnam.

A former participant in government policy-making, himself an author of a book about Vietnam, noted:

I've learned facts from some books and I've gotten arguments from other books, but I don't think they evolved my view. Meditation and conversation, I suppose, did. I mean, I can recall incidents as well as talking with people, but I can't recall reading a book and getting a sense of great revelation. Most of them weren't saying much about policy.

In short, most respondents read books, as one highly influential New York free-lance intellectual admitted, "to back up [their] prejudice." Only one percent was classified as having been converted or totally influenced by a book, although most claimed to have read at least part of a book. Nonetheless, as compendia of material previously published as articles – and almost all the books included previously published material – the books were important guides to the sources of influence on intellectuals. The list of leading books will be taken up when we discuss influential intellectuals.

If the *New York Times* gave respondents the facts which led many to change their minds, and books helped bolster their arguments, articles in the leading intellectual journals were crucial for 16 percent in supplying them with an opinion. Here is an example of a person whom we classified as a middle switcher:

I have an open mind, but I don't think that any article now is going to persuade me that our policy is right. But by the same token, I think that the articles that I've read have had a large part in changing my views back there in '64 or '65.

Not only were people who changed their minds affected by journals, but journals also affected those – such as this pacifist, non–Cold Warrior who had always been opposed to the war – who began to form strong opinions only around the time of the Tonkin Gulf resolution:

> I'm not that sharp to figure out things for myself, and so there were people who helped me to "tune in" fairly early, like T. R. B. in the *New Republic* and like I. F. Stone [in his own journal].

Everyone, of course, had read some article on Vietnam, but journal articles were the major influence on 16 percent of the sample. Again, which were the leading journals is a matter to be taken up when we evaluate the opinions of the leading intellectuals.

The last medium, TV, is difficult to evaluate for a group such as ours. To be truthful, we did not include it in our standard probes. Even so, about 10 percent of the sample mentioned the impact of TV pictures of the war and five specifically mentioned Huntley and Brinkley. Given the reluctance of intellectuals who grew up in a pretelevision era to acknowledge its existence, much less to admit its impact on their own thinking, we are not surprised that only one person formed his opinion about Vietnam because of what he saw on television. But then, this figure just about matches the number who said they were convinced by reading books.

Influences which are more direct upon a person are, curiously, a bit more difficult to evaluate. The largest category is direct personal experience. As an editor who had traveled extensively in Southeast Asia remarked, "Neither magazine articles nor books have as much impact on my thinking as the actual experiences I had in connection with the events." Our classification of "actual experiences" is quite broad. It includes those who actually were in Vietnam, those who were in special positions to receive information outside of the usual formal channels, such as those in decision-making positions in the government and those who were editors of journals which possessed a large news-gathering operation, and those who in a professional way did their own research on Vietnam and thus were direct recipients of information they, themselves, set out to find. Frequently a person was subject to all of these influences.

A considerable number of our respondents, 15 percent, had

actually been in Southeast Asia in the last ten years, on either official or privately sponsored trips. There is, of course, no guarantee that this experience alone was responsible for their opinion change. More dramatic than most, however, is this interview with a correspondent who was an early switcher:

> I was an X-year-old kid sent out there by ——, kind of spilled out of the helicopter trying to scratch around and find out what's going on, and yet, the essential reporting in those days stands up pretty well. I was there as a correspondent in 196– and I just didn't think that any of the things that we said that we were doing we were actually accomplishing. . . . It was such a personal experience that the only people that could influence you were really only your colleagues at the time. . . . But I just think the whole thing is a great, extraordinary study in futility. . . . And I hate the goddam war.

Less dramatic, but in a way equally personal, is the situation of an editor who sits in his office assembling information from all kinds of sources. After we completed an interview with the editor of a liberal news weekly and after we sent him a transcript of the interview (a somewhat unusual procedure, but one which we tried to comply with, when requested), we received the following letter, from which we have been given permission to quote:

> A person in my position is in touch with far more sources of information on any particular subject than he could possibly recall in an interview of this sort. I confer with people all day long. . . . Then I take a look at manuscripts that are submitted. One naturally absorbs information from manuscripts which may have been submitted under these circumstances, even though the manuscript may have been returned. But no editor, I am sure, keeps a diary of the persons that he had been in contact with on any particular subject, and much less could he specify even if he wanted to, the extent to which his thinking has been influenced by such contact.

The editor then went on to list the enormous variety of sources with which he is in regular contact. Though most editors we interviewed were not in the position of putting out a news weekly, most of them were in fact mainly influenced by their own reporters and writers, many of whom, upon returning from Vietnam, first gave a private account to the editor. And although most editors try to glance at what their competition is putting out, the only source of information in depth they have much time for is generally that provided by their own journal.

Government officials or government advisers are in a similar position, getting information from special and private sources, though in our data professors who were also advisers to government seemed to utilize a somewhat wider and less parochial set of channels. Yet frequently these channels are supplemented by on-the-spot observations which allow for some independent evaluations of reports. A government adviser reported:

> In 1961 when things got worse in Saigon, I went there and I've been skeptical about our operations ever since. . . . I thought the Taylor-Rostow report a great piece of nonsense.

Finally, included in this category of direct experience are persons whose information and evaluations come to them not from editorial, government or special research channels, but from party discussion and other party sources. Such persons also receive information from a variety of indefinable but special and privileged sources.

All in all, this category of personal experience and special sources is the most frequent category, as one might expect from a group of opinion-makers. One third of the sample initially made up their minds about Vietnam as a result of such sources.

Direct personal confrontation with friends or acquaintances was responsible for forming the opinion of about 20 percent of our sample. This is a minimal estimate, since most intellectuals are reluctant to admit the force of another person's argument. A member of the New York intellectual "family" gave this personal testament on how he switched from support of the United States to opposition to its policies.

> It was around 1963. . . . I was sitting in the home of some friends. . . . We were talking . . . about the Vietnam thing and I took the position, "Well, after all, aggression, and one has to make certain, etc., etc. . . ." And [the wife] laid into me like a ton of bricks and said I hadn't done my homework, that I really didn't know what it was all

about. . . . I felt very chastened, indeed, and . . . that is the first time I systematically began to read about it.

As in most cases of personal influence among intellectuals, the impact of another person was then buttressed by an active search for documentary evidence. But the scales of evaluation had been tipped.

The final category is a residual one in the sense that it represents diffuse personal encounters and experiences with an entire cultural atmosphere composed of teach-ins, marches, rallies, demonstrations, and countless unremembered conversations. As one leading critic, a central member of New York intellectual circles, who was a late switcher, put it, "I came to my position fairly late. Well, I would put it at 1966 . . . No particular events. A sense of the culture – quite abstract." Thirteen percent fell into this group.

To sum up, despite the fact that intellectuals are by definition omnivorous consumers of the printed page and our respondents themselves are responsible for a very large proportion of the avalanche of printed material on Vietnam and Southeast Asia, the printed word was chiefly responsible for the opinions on Vietnam of only 35 percent of our sample. The rest, although everyone read a great deal about the war, formed their opinions mainly as a result of their social situation and interactions which resulted from that situation.

Each decision about opposing the war had a characteristic source of influence. Those who had always opposed United States policy but who became active in their opposition fairly late relied heavily on the media – two thirds did so – and they were the only group that did so. Thirteen were mainly affected by newspapers – the *New York Times*. This group clearly had their latent predispositions activated by the news. They came to their position later than those who had opposed the war earlier because they did not have the firsthand experience of the majority (14) of the early opposers. Journals were the other key source for 6 of the early opposers simply because if one did not have firsthand experience with the problem there was no other way than through articles, in the early days, to find out about it in any depth.

Not surprisingly, both the early switchers and those who were always for the United States policy were also mainly influenced by their own direct experiences (4 of the former and 7 of the latter). For both groups, their opinion meant running against the tide. The early switchers were mainly Cold Warriors, most of whose friends still favored supporting the Diem government. And those who consistently supported the United States policy have had to buck the overwhelming majority of their fellow intellectuals.

Middle switchers were essentially affected by two sources. First, the general social environment, especially the atmosphere of university campuses with the teach-ins of spring, 1965 was most influential for seven of them. Second, five were impressed by journal articles that had begun to appear with greater frequency – by material in the *New York Review of Books, Commentary* and the *New Republic.*

The most characteristic influence upon the late switchers was direct personal discussions. Half (six) finally changed their minds in this somewhat arm-twisting way. If one is long a holdout, then it takes a powerful effect to make one change. Which brings us to a more detailed examination of the structure of discussion on Vietnam.

Circles of Discussion

The pattern of discussions of the war in Vietnam was very different from the pattern of intellectual circles shown in Chapter 3. Figure 2 shows essentially a central core made up of Vietnam experts and their associates, each of whom checked with each other, and a penumbra possessed of relatively few independent centers of interaction.* Persons in the outer circles

* The computer worked on the answers to the following set of questions:
When did you become especially concerned about Vietnam (if name of person was given)?
What individuals made you especially concerned about America's involvement at this time (if the name of a conversation partner was given)?
Since your initial concern, have you talked with any particular person or persons who have had a special impact on your thinking?
In the last six months, with whom have you most often discussed the Vietnam War in depth?

tended to check back with those in the central core, either directly or via one or two associates. Curiously, the density of contact for the permanent circles and for the Vietnam discussion circles is about the same. It is the way that they are structured which differs. Two possible exceptions are the radical "Cambridge" circle and the more conservative (on Vietnam) *Dissent* crowd. This is quite different from the pattern of the more permanent intellectuals' circles, which showed two main overlapping circles with several smaller ones barely connected.

The Vietnam issue had been around a long time, and the patterns shown are the net result of many years. One respondent noted that "the Vietnam war has not been subject to intellectual debate for a long time." Several remarked that it was hard to find people either on campuses or in New York with whom to debate the issue. In fact, the staleness of the issue worked against our attempts to get a picture of the discussion patterns.[3]

> One of the great booby traps of all intellectual operations [is that] you just move among people who agree with you. . . . I do know a couple of well-known intellectuals with whom I disagree. I try to keep in touch with them so that I can fight with them more.

The pattern therefore includes persons of differing opinion, to the extent that strong differences still existed among intellectuals in 1970. But the important fact about the discussion pattern was that the inner core started the discussion a number of years ago and was still in 1970 regarded as the source of knowledge for almost all intellectuals, regardless of political persuasion. The lines from the core outward, however, show a remarkable degree of participation and in this respect the vast majority of elite intellectuals served as opinion-makers for each other.

Some characteristics of the inner core are important for understanding how the intellectual community shifted. First, there was a strong relationship between centrality in intellectual life in general and centrality in the Vietnam discussion. Nine of the ten center circle members (see Chapter 2) were also in this Vietnam inner core, but only one third of the members of other

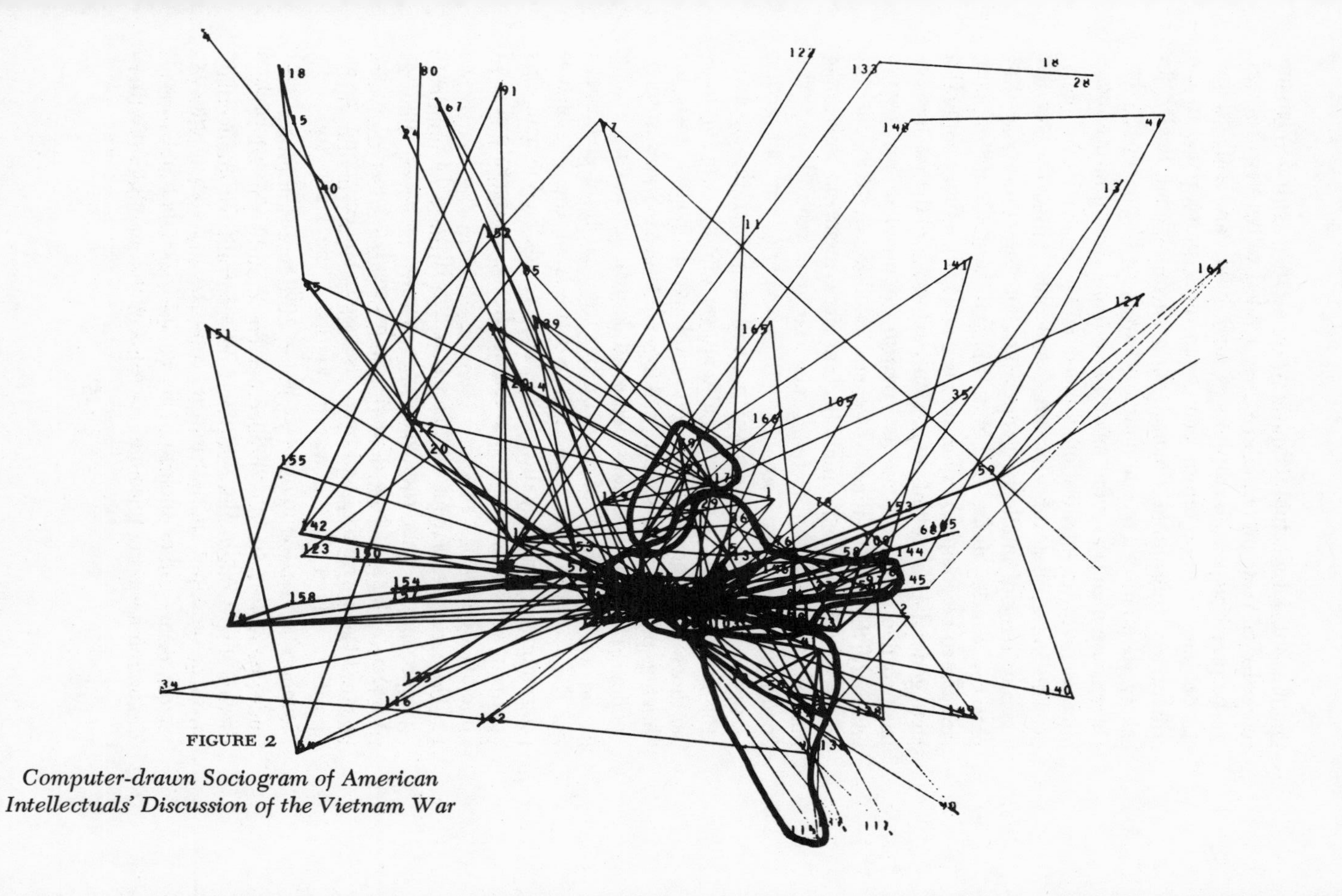

FIGURE 2

Computer-drawn Sociogram of American Intellectuals' Discussion of the Vietnam War

permanent intellectual cliques were in this central Vietnam core. Inner Vietnam circle members tended to be New Yorkers, to be prestigious, to be authors of more books and articles, and to be Jews. Even more of them than other participants in the discussion considered themselves intellectual, and finally, to show their special cosmopolitan character, I can report that inner core members were much more likely than noncore members to favor the legalization of marijuana.

The inner circle of Vietnam discussants, however, was not merely a set of intellectual superstars. On the contrary, except for the center circle members, the Vietnam discussion core tended to be composed of different people from those named as top general intellectual discussion partners. Of the 20 persons named five or more times as Vietnam discussants, only 3 were named five or more times as general intellectual circle stars; of those who are named top intellectual circle members, only 3 are in the top 20 among Vietnam stars. To give some idea of who composes the inner core of Vietnam experts, without giving away their names,[4] we can say that half of those in the top 20 in the Vietnam discussion group also appear in our list of top book authors and another five were intellectuals well known for having participated in the making of United States foreign policy. These top leaders were all known as having strong minds on the subject of Vietnam but were far from unanimous in their stand. Thus, although two thirds of the intellectual community eventually did change their minds on Vietnam, this pattern of discussion shows that radicals, conservatives and liberals all had their say.

The more permanent intellectual circles discussed in Chapter 3 represent the net resolution of a number of issues such as Vietnam. The structure of these circles tends somewhat to lag behind the way issues had been discussed. The consequences of positions taken on the war in Vietnam, the New York City teachers' strike, the militancies of the 1960's, and so on, are just now beginning to be felt in a serious way in the intellectual community. While the discussion of issues is going on, there will be gossipy articles in the *New York Times Magazine*, *Esquire*, and other such journals to the effect that so-and-so is no longer speaking to so-and-so. All of which may be quite true.

But the impact of a number of so-and-so's not speaking to so-and-so, and the consequences of new alignments may not actually acquire solidity for some time. When we interviewed intellectuals in 1970, the circles we found were in part a resolution of the issues of the sixties, but much more a reflection of the issues of the forties and fifties. So when it comes to the Vietnam issue, the circles we described as "radical," "literary," "central," and "social science–literary" only partly reflect the outcome of the debate.

The radical circle is, of course, different from any other circle in its stance on Vietnam. Two thirds of the radicals wanted to get out now, three quarters were concerned about the war before 1963, 100 percent had written on the war, and all were always against the war. (For details, see Table 15.)

All this follows, of course, from the fact that in the late fifties almost all of them were opposed to the Cold War. In the other circles, only one third or so wanted to get out right away, and the social science–literary circle included 20 percent who wished to prevent a Communist takeover, making them the circle with the greatest number of conservatives. And half the members of the other circles actually switched their minds on Vietnam, being at some point in favor of American intervention. Curiously, fewer of the isolated changed their minds – perhaps because they were subjected to less social pressure. The circle whose members most changed their minds on Vietnam was the literary circle, whom we had earlier seen as most affected by the so-called "radicalization" of the sixties. Three quarters of this circle favored containment in the late fifties and though only one third favored immediate withdrawal at the time of the interview, none wished to prevent a Communist takeover. What is more, their interest was sparked early: like the radicals, three quarters became concerned about Vietnam before 1963. All in all, this represents a considerable change.

To conclude: the pattern of an inner core of experts and an outer set of interested intellectuals may well be typical for the discussion of any given issue within the intellectual community. But Vietnam is unique in that it had remained an unresolved issue of daily front-page character for more than ten years. This has had two important consequences. First, the long time the

TABLE 15

Intellectual Circles and Vietnam
(In Percentages)

	RADICAL (N = 6)	LITERARY (N = 10)	CENTER (N = 10)	SOCIAL SCIENCE–LITERARY ONLY (N = 36)	NON-MEMBERS (N = 48)
Cold War position before the late 1950's					
Opposed	83	0	0	22	30
Moderate	17	25	40	33	12
For containment	0	75	60	17	42
Hawk	0	0	0	28	16
Vietnam position					
Get out now	67	33	40	33	45
Favor coalition	33	67	50	47	43
Prevent takeover	0	0	10	19	13
Year became concerned					
Before 1963	75	75	22	36	36
After 1963	25	25	78	64	64
Wrote on Vietnam					
Yes	100	56	70	66	46
No	0	44	30	34	54
Decision type					
Always for the U.S. position	0	0	11	14	11
Once for, changed mind	0	51	55	45	36
Always against U.S. position	100	50	33	41	53

issue had remained before the intellectual community despite the community's usual tendency quickly to move on to the next "hot" issue reduced the vigor of discussion, and it contributed, as in the case of our other long-term problem, race relations, to an unusual lack of clarity and imagination in the proposals

TABLE 16

Intellectuals' Attitudes on Vietnam War and Attitudes toward the Cold War before 1960

ATTITUDE ABOUT VIETNAM INTERVENTION	ATTITUDE TOWARD COLD WAR BEFORE 1960		
	"STALWARTS," STRONGLY SUPPORTIVE[a] (POSITIONS 1, 2) (N = 51)	"CRITICAL SUPPORT," MODERATELY SUPPORTIVE (POSITIONS 3, 4) (N = 23)	"RECALCITRANTS" AND OTHERS STRONGLY OPPOSED (POSITIONS 5, 6) (N = 26)
Always favored U.S. policy	25	0	0
Switched attitude			
Late	18	17	0
Middle	34	17	0
Early	6	13	0
Opposed U.S. policy			
Late	8	22	46
Early	8	31	54

[a] See Chapter 4, pp. 105–112 for definitions of these terms.

presented for solving the problem. Second, the jumbled pattern of discussion which has cut across the lines of intellectual circles has undoubtedly led to the general weakening of existing intellectual circles. Intellectual circles usually reflect the way the *last* set of issues was resolved. When a new, hot, perennial issue remains unresolved and floating, as it were, the fabric of intellectual life suffers severe damage. All of this means that before we examine more closely the characteristics and opinions of leaders and influentials in the intellectual community on the topic of Vietnam we shall have to turn to an examination of the general style of argument on that issue.

6

The Triumph of Pragmatism

The elite American intellectuals no doubt affected the progress of the war in Southeast Asia. They provided a rationale for extricating the United States from Vietnam so that by 1966 at least – perhaps by 1965 – when almost all the intellectuals who were going to change their minds had already done so, it became chic in some circles to oppose the war. Opposition was not confined to radicals, leftists, long-haired youth or cranky, maverick Senators. A groundswell had begun which would eventually create a climate such that no responsible American leader could possibly claim the war a good thing – it became merely a case of how to get ourselves out of it. Events themselves were important, as most of our sample of leading intellectuals pointed out. But having a respectable rationale for action is always important, and if intellectuals have any function at all, this is one of their chief ones.

The key is respectability, of course. Because the war did not radicalize the American intellectual elite, most were able to provide respectable arguments. Respectability, in the American tradition, means practicality. The overwhelming majority of American intellectuals took a practical line – we got sucked into the war through a series of step-by-step mistakes. The war itself was a mistake because it could not be won. This summary of the position of most American leading intellectuals may not square with the reader's own impression, nor does it coincide with the views many intellectuals have of themselves. Rightly or wrongly, happily or unhappily, our view is correct and the reader is invited to follow the argument in detail. In the course of the analysis he will also find the source of his mistaken impressions.

How We Got Involved in Vietnam

The spectacle of Americans fighting a land war on an Asian peninsula only ten years after the Korean stalemate was so bizarre that all official explanations for the war, and many private ones as well, had to account for why Americans became involved. A rationale for or against the war was often based on a historical foundation. Thus, although we did not specifically ask most respondents for it, more than half the intellectuals interviewed volunteered some historical account. What kind of intellectual offered an historical explanation?

Although history is a cornerstone of Marxist analysis, and revisionist views of the Cold War have become quite fashionable in some quarters, most of the leading intellectuals who invoked historical explanations of the war in Southeast Asia were not radical to begin with, nor were they radicalized by the war. Rather, those with historical explanations were mainly those whose position on the war was equivocal or who were Vietnam experts. Some who blamed Cold War ideology for the war in Vietnam more fundamentally questioned American policies; these, however, were a distinct minority who had always held contrary views.

By an equivocal position we mean one favoring a reduced United States involvement in Vietnam while at the same time pressing toward some coalition government. Half of those taking this position attempted to explain why or how we had gotten into Vietnam. Only one fourth of those advocating some other position on Vietnam talked about the history of our involvement: those with a firm demand for immediate withdrawal apparently had less to explain or did not care why we were there as long as we left right away; and those who still wanted the United States to prevent a Communist takeover were more interested in Cold War geopolitics in general than in the specific details of what even they admitted to be a misadventure.

History is also the province of the expert. Those whom we classified as having been most influenced on the war by their own direct experiences or research were considerably more likely than others to discuss the history of the Vietnam War.

Expertise and equivocalness together explain a great deal: almost 90 percent of those who volunteered a historical explanation, as compared with 33 percent of those who did not, either took the coalition position on Vietnam or had special expertise.

Exactly half of those who discussed the history of the war mentioned one or another aspect of the Cold War. In turn, the Cold War, as an ideological battle, was seen as the major cause of the war in Vietnam by half of those who invoked the Cold War. Ideology explained what might otherwise have been a total mystery:

> I wonder what really keeps this war going on. This is a war historians will wring their hands over. I think in the end what keeps it going is this fundamental fear of Communism. Not that individual nations or one shouldn't be afraid of Russia, for example. Or that one shouldn't be afraid or wary of many elements of what is called Communism, though Communism is such a disparate thing. . . .

"Fear of Communism" not only kept the war going, but was the very reason we entered in the first place. Almost all who took this position denied traditional Marxist economic explanations for American involvement, emphasizing that ideology was responsible, not economic imperialism. For example, American political expansion

> links the American concept of enterprise with Christianity and with a religious belief. So I see what happened in Southeast Asia as part of the Cold War. And I don't believe that the Cold War was fought

solely to preserve American economic interests. I think that is much too simple a notion. I think there really were ideological commitments on the part of many Americans so that the Cold War lasted a long time after the ideological commitments had ceased to be terribly important.

Several democratic socialists directly attacked "leftist" views of the war. "I think we got in there for ideological reasons and those on the Left who think we got in there for materialistic reasons are basically wrong." Only one respondent endorsed the radical revisionist idea that the war is a result of basic American economic structures.[1]

Among those who cited the Cold War as the major cause of United States involvement in Vietnam were several who basically approved of the Cold War. For example, an editor took a very personal view of the matter.

> I used to support the war. . . . I originally thought the war was worth winning for the same reason that the war in Korea was worth winning. . . . I don't bring to it [his present opposition to the war] this enormous indignation and violence of feeling . . . the same people who took us into World War II took us into Korea for much the same motives: stop dictatorships, draw the line against tyranny, etc., and to preserve in various parts of the world the balance of power. The fact that World War II and Korea worked out satisfactorily from our point of view and Vietnam is not working out satisfactorily, that's just life. Some wars you win, and some you lose.

The Cold War was seen by some not as an ideological issue but as a series of responses by the United States to threats to national power.

"It [Vietnam] entirely parallels the problems we faced in Korea and Berlin and in Greece and in Iran of trying to establish and maintain the balance of power with the Soviet Union. . . ."

Finally, there were those who saw Cold War ideology or geopolitics as a cause of the American involvement but who also considered that a series of blunders rather than firm policy or foresight led us into the war. Typical of the mistake-plus-Cold-War-policy analysts is the following point of view:

We stumbled and fell into it on all levels. I don't think anybody wanted the degree of involvement that finally happened. Everybody felt the prisoner of what the guy before had done. . . . Somewhere there was left over in the minds of the whole series of administrations the Cold War notion that the basic shape of the world is the fight between the forces of good, mainly us, and the force of evil, mainly them – the Communists, and that Communism amounted to a series of conspiracies directed from the Soviet Union ultimately against the people of the world, and we were the deliverers. All this collapsed utterly in Vietnam. . . . One of the great casualties of this war is the Cold War mythology. Nobody believes in it anymore.

That nobody believes in the Cold War mythology anymore was belied by the following respondent, who agreed that we entered Vietnam as part of Cold War policy, that we blundered there, but who also felt that the Cold War was nevertheless very much part of the current world scene.

Given the conditions that were prevailing in the early 60's . . . and given the commitments previously made, we really had no choice but to send troops in.

In a way, saying that the Cold War was responsible for our entry into Vietnam is to say that given the premises from which the United States was operating – and whether they were right or wrong premises is another matter – the Vietnam involvement was a natural, logical, or reasonable outcome, even if we got sucked in step-by-step. One quarter of the intellectuals who discussed the history of our involvement gave this step-by-step explanation without directly mentioning the Cold War.

All right. I see very clearly how we came to be there. . . . I, too, recognize that having got where we are by relatively gradual stages, each might have made some fair measure of logical sense.

Almost all of the nine "step-by-step" theorists either directly blamed the French, or our policy toward the French, or the "vacuum" they left as the major cause of our drift into war. Perhaps the main point of the step-by-step theorists was the nonrational character of American involvement. "It seemed to me to be influenced by considerations that did not have much to do with American interests. It was influenced by European considerations . . . [and by] a morbid fear . . . on the part of policy makers. . . . They did not want to be in a position where their political opponents could say, 'You did nothing. You let South Vietnam go Communist.' "

Clearly, a step-by-step explanation does not rule out the possibility that those responsible for taking those steps were inadequate, fearful, or foolish. For that matter, several of the advocates of an ideological explanation also blamed American leadership for having fallen prey to Cold War thinking. But ten of the 48 intellectuals talking about history directly placed the major blame for American involvement on the character of American leadership or upon specific leaders. Lyndon Johnson obviously comes in for his share of knocks. For example, a political scientist said, "We overreacted [to the Gulf of Tonkin incident] because of the desire on the part of Johnson or his advisers to get involved over there, to have an excuse for bombing North Vietnam." But considering the vilification heaped upon Johnson by some intellectuals who dissented from his position on the war, and the fact that most of our sample became opposed to the war during his tenure in office, it is remarkable that only three intellectuals placed the blame for the war mainly upon him. And only a few others mentioned him in the course of comments about Cold War ideology or the step-by-step unfolding of events. Perhaps distance makes the heart grow softer. And as one might expect from the support he received from intellectuals, only several respondents attacked Kennedy, and then frequently on psychological rather than political grounds — "his reactions that no one was going to put him down, and beefing up around women was an outgrowth of weakness, an element in the American male character."

Most of the intellectuals who blamed people as the cause of the war mentioned John Foster Dulles (almost all of those who spoke of the French period referred to him), but even more respondents blamed advisers or the system in general. One respondent, an expert on foreign policy, while noting that the war was "a mistake that flowed from a misconception of our role in the world, our conception of the fact that wherever something called Communism reared its head, whether there were only three communists involved as in the Dominican Republic, we had to charge in with the Marines and set things right," said that "the basic flaw goes back again to the military dominance over policy making. . . . They came out of World War II with a halo around them and the Cold War had scared us out of our wits under Truman, so that the military became the infallible experts on everything."

The theme of "they" did not know what they were doing is frequently repeated.

> I don't believe they've ever sat down and figured out what they think the problem is in stopping, let's say, the Communist menace abroad. . . . In other words, the troops being sent there [Vietnam] were not part of an overall policy that could be understood and that could make any sense.

Finally, not even the intellectuals themselves escape being held responsible for the war. A leading intellectual who was associated with the government blames his own kind:

> Vietnam policy is primarily the policy of eastern intellectual elites who have failed very badly with that policy and have now disavowed it because they have had a change of administration which enabled them to suggest that it somehow is military and somehow Republican, etc., but primarily this was liberal democratic elites who started the war and prosecuted it to the point where they failed.

An art critic despairs of anyone's, or any particular President's, being responsible and hence, in a sense, argues that each one bears a share of the blame.

> Everybody says, "He did it." Johnson says, "I got it from Kennedy." Now Nixon says, "I got it from Johnson." Each time there's a claim that we're stuck with a situation and that our policy is the only possible one within the particular situation and that we can't abandon our allies, as they put it, to be massacred. . . . I mean, everything flows from everything else in a kind of moronic logic that begins with the assumption that nobody's responsible.

In short, there were basically three approaches to an informal "historiography" of American involvement in the war in Vietnam: Half (24) of the intellectuals said the Cold War was in some way responsible; 12, or one quarter, gave a "step-by-step" explanation with the policy of aiding the French in the early 1950's taking the lion's share of the blame; and another 10 found fault with the decision-makers. Two came up with other kinds of answers.

All of the three approaches to Vietnam history taken by those who chose to offer an historical explanation are essentially liberal-realist explanations.[2] American ideology, and the application of the ideology, the workings of Southeast Asia foreign policy, or the poor character of the people making the decisions, or all of these may have been responsible. American economic imperialism was not considered as a cause of involvement, as some of the radical critics might have it. To be sure, only one radical chose to talk on the specific history of Vietnam. The other radical critics in our sample spoke more generally about the character of the United States. The details of Southeast Asia history are, to them, somewhat beside the point.

Despite their essential agreement on one level, proponents of the three types of explanation differed from each other in several ways. To begin with, those offering Cold War explanations are younger than the others. Four fifths are under fifty-five, as compared with two fifths of each of the other groups. The Cold War as a reason for action is perhaps more salient to them since their entire adult life was spent under its shadow. As one might expect, those taking the step-by-step position were also more moderate in their demands for withdrawal – only one advocated immediate and unconditional withdrawal from Vietnam. But their moderate approach does not mean they are passionless. Curiously, the coolest discussants of the Vietnam issue are those who blame the decision-makers. Only two of them were highly emotional when speaking about the war in Vietnam. This suggests again that their dissatisfaction with the decision makers is of a very different kind from the anti-Johnson hysteria of the middle 1960's.

Earlier we said that both opponents and proponents of the Cold War cited it as a cause for American involvement. When we separate out the opponents of the Cold War who saw its ideology as basically responsible for our entrance into Vietnam, then a distinctive group emerges: five of the seven are democratic socialists or currently leftists of one kind or another.

The views of intellectuals on Vietnam historiography given here are drastically different from those publicly available, for people holding different positions on the history of the war were not equally prone to publicize them. The impression that the war in Vietnam has radicalized intellectuals and that the intellectual community took a radical stance on the war is caused by the distribution of published opinion. For it is leftists who opposed the Cold War and who saw its ideology as chiefly responsible for the United States involvement in the war, and proponents of this position were far more likely than others to write about their views. Further, a step-by-step position is essentially a conservative one, for the only lessons it offers for the future lie in the direction of caution at the least, and an overhaul of intelligence and top-level decision-making machinery at the most. Only two of the step-by-step theorists had put their opinions in print at the time of our study.

Why American Policy in Vietnam Was Wrong

Three things were wrong with American policy in Vietnam, according to the 99 members of the American intellectual elite in our sample who offered reasons for our lack of success: it made no sense pragmatically, and was wrong morally and ideologically.[3] Now, one might assume intellectuals tend to be either moralists or ideologists. Not so American intellectuals. They are firmly steeped in the pragmatic tradition, having renounced whatever revolutionary or ideological tendencies they had acquired in the thirties and forties. The Vietnam War did not bring a return to ideology; rather, the overwhelming majority of leading intellectuals were pragmatic about the war; moralists, of course, have never had a chance in twentieth century America. In the present case pragmatism did not mean that the entire American system was at fault and did not work, rather, that its foreign policy was at best misguided and at worst stupid. The war was simply not in the interests of this country.

To document these views of the way intellectuals reasoned about the war, it is necessary first to explain what we mean by pragmatic, ideological, and moral. Pragmatism denies that means can be separately evaluated from ends. There are no absolute ends or goals because the very search for such ends may involve people in undesirable actions or activities. Only what "works" now is valuable. In contrast, an ideology claims to be a more or less logically ordered, closed system of beliefs about specific ends, goals or values such as "democracy," "socialism," "Communism," or even something less overarching, such as "Cold War."[4] Courses of action are "scientifically" evaluated in terms of their efficacy in reaching their ends. In one sense, because it is a system of ideas dealing with human preferences, pragmatism, too, is an ideology. But the system deals with how to go about solving problems, that is, it is about method, not substance, and so pragmatists claim that they have no ideology. In their opinion, human problems can best be solved through the application of technology and systems theory to human affairs; they need not be reserved for mechanical engineering. Hence, the "End of Ideology" was proclaimed in the 1950's. Ideologists feel that ends cannot be random, but must conform to some kind of system. Naturally, whatever system they have chosen can be logically defended against all others, which can and must be shown to be false.

In contrast to both pragmatism and ideology, moralism claims more instant understanding and evaluation. A moralist needs no evidence other than his senses to judge something right or wrong, and no elaborate scientific calculus to ascertain what the proper course of action should be. Like the ideologist, the moralist has his preconceived set of values, but unlike the ideologist he does not necessarily claim that the values are systematically or logically arranged. The pragmatist, along with the moralist, denies that values can be systematically arranged in logically neat hierarchies, but, like the ideologist, the pragmatist insists that proper courses of action can be scientifically and logically evaluated.[5]

Finally, there is the idealist position, one taken by a group of

New York intellectuals in talking about the Cold War but one which, as we shall see, they were forced to abandon when it came to Vietnam. Idealists, like ideologists, are much interested in ultimate values, are willing to arrange them into some system, but do not feel that ends can rationally be evaluated. In this sense, the ideals are "ultimates" though unlike moralists, idealists do have a "system."

All this may seem very abstract, but the effect of the war in Vietnam on American intellectuals is deeply tied to whether they took a pragmatic, an ideological, or a moral point of view of the war. Despite some claims to the contrary, or perhaps some wishful thinking by leftists, the war has not encouraged an important shift toward ideological thinking on the part of leading intellectuals. And, despite the activity of some prominent churchmen, the war has also not spurred a moral reevaluation of American life among the majority of intellectuals we interviewed and certainly has not led to any significant increase in the respectability of moral thinking among American intellectuals. The war in Vietnam was opposed by the overwhelming majority of the leading American intellectuals in our sample simply and solely because it did not "work." Naturally, those who already opposed the Cold War were more likely to oppose the war in Vietnam on ideological or moral grounds and to oppose it much earlier than others. But the fact that in 1970 the war could be opposed on pragmatic grounds alone by many who had not strongly opposed the Cold War is crucial to understanding the 1960's.

Let us look more closely at the version of American pragmatism which swung opinion on the war in Vietnam. From the very start, it was possible to oppose the war from a strict pragmatic point of view, as this interview with a very influential foreign policy expert demonstrates:

> I wrote before on this, not with regard to Vietnam in particular, but when I travelled to Asia for the first time in 1955 I came back with the impression that we had bitten off more than we could chew. And I wrote a series of articles for —— Magazine. . . . [The war in Vietnam] was based upon a completely wrong assessment of reality – a wrong assessment of our interests, of our power, and of the interests and the power on the other side. . . . You see, my position has al-

ways been non-ideological – simply practical. . . . If you want another metaphor, I'm in the position of a businessman: "Don't open that store. You will go broke." It doesn't mean that this guy is anti-capitalist, or something. He simply looks at the situation and arrives at the conclusion that this enterprise cannot succeed.

This is the classic "limits of power" position expressed by 46 of the 78 intellectuals who gave pragmatic reasons for opposing the war.

In his introduction to the *Commentary* symposium in 1967 on "Liberal Anti-Communism Revisited," Podhoretz argued that the 21 participants reflected

the extent to which the liberal and intellectual communities had lost the faith they momentarily had at the height of the Cold War in the possibility that the United States, the main bulwark against Stalinism, could act as a relative force for the good in international affairs. Not everyone in this symposium entertained that faith in the 1950's [indeed they did not – see above, Chapter 4], but most to some degree did, and not many do any longer.[6]

This loss of "faith" might be termed a shift away from idealism to pragmatism.[7] The concern about this loss of faith is reflected in this following almost quintessential pragmatic position.

I would have been in favor of governmental help to Spain. . . . However, as I have also pointed out, that though morally justifiable, sometimes it is not prudent . . . for the situation is such that it is strategically impossible to win such a war. Or it may weaken the democratic potential as a whole in its resistance to totalitarianism. So the mere fact that morally I am committed to help any freedom-loving force to resist foreign totalitarian aggression, I cannot intelligently and in good conscience say you must always do it. *Discretion has to be added here to principle* [Italics ours]. . . . Now I would regard that whole Southeastern Asian situation as one in which the possible gains hardly justify the likely losses. . . .

I've actually come to the conclusion that it was a lost cause, and my only question is how to get out with the least amount of human suffering. . . . I must confess the moral problem involves me even more than the so-called domino theory. I'm prepared to believe that if the

United States acted in such a way [immediate withdrawal] there would be a moral calamity and consequence that would hurt the United States.

Pragmatism may mean testing truth by practical consequences of belief, but this testing clearly involves a complex calculus. For example, this respondent obliquely refers to the "blood-bath" argument, as did many pragmatists. To the advocates of a pragmatic approach, therefore, the essential point is not that a pragmatic evaluation of United States policy in Vietnam is amoral, but that it consists of weighing sets of *contradictory* morals. In this calculation, there can be no absolute morality such as "war is bad" or "killing is immoral"; this is precisely what most offends those we call moralists. A pragmatist calculates whether some moral good should be made subservient to some other moral good, though, to be sure, some pragmatists specifically denied that any calculation of morality was involved.

Whatever it may pretend to be, pragmatics is therefore not solely a neutral, objective estimate of consequences, for it is not possible to evaluate what works without considering the ends in view. Most pragmatists have a clear political point of view, as do the respondents just quoted. Basically, they favored the ideological premises of the Cold War and did not feel any need to question them. Some supported these premises from a position of faith in America. Most, though not all, would have preferred the United States to dispose, without too much fuss and bother, of what was to them an evident danger in Southeast Asia. In fact, most of them were not disturbed by Kennedy's sending of advisers, and were middle or even late switchers. As one man put it, "The sending of military advisers . . . didn't seem much of an issue one way or the other. . . ." The basic problem was that it did not work. As this respondent frankly admitted, "I must confess that when the troops were first sent in, I defended that which I came, a few months later, to regard as an error . . . because we kept sending in more troops and it was obvious that once we started this we set the war machine in motion in a fatal way." The war was "fatal," not because war per se is wrong or it was wrong to attempt to hold back Communism, but because the war was prolonged and had serious domestic consequences.

Intellectuals who objected to American Vietnam policies on pragmatic grounds almost always said, "We can't win." Curiously, most of those taking this position found it unnecessary to explain *why* the United States could not win, or *why,* after the successes of World War II and to some extent those of Korea, we could not beat a second-rate Asian power. To most Americans and even most policymakers this impotence came as a shock. Most of our respondents thought the reasons for failure were obvious. Those few who did explain the why's of our inability to win were generally either less politically sophisticated or, paradoxically, were government advisers who felt obligated to show off their inside knowledge. And very few respondents, indeed, went into any military details. These are matters either beyond their interest or their self-perceived competence.

The pragmatic approach is not confined to an evaluation of possible American successes or failures in Southeast Asia. About one third of the respondents discussed the pragmatic consequences of American policy in Vietnam in terms of their overall consequences for the United States itself or for our relations with the rest of the world.

The argument closest to the "we-can't-win-in-Vietnam" position was one which explained that we could not win, not because of the nature of American power overseas, but because of the lack of support for the exercise of that power at home and the consequent political polarization of the United States itself.

> The amount of internal distress over the issue indicates that [choices] three and four [prevent a Communist takeover or actually try to "win"] are very risky. . . . When there's a real war, when the Japanese start dropping the bombs, then all Americans will go all out and support the government, but they will not be drafted for a police action.

An editorial writer explained:

> I'm broadly anti-Communist and because I thought America had the military strength that it could block the Communists, I supported

Johnson in '65 to '68. I used to write on behalf of the war. . . . But by '67–'68, that winter, I began to feel that whatever we were hoping to gain in Southeast Asia would be more than offset by what we were losing in the United States. The tone of public opinion was getting so hysterical, it was creating a serious problem for the United States – the bulwark, the linchpin of the whole free world coalition.

In this case, domestic unrest affects our foreign relations because of the possible weakened condition of the United States. Domestic unrest, of course, was not limited to the students or the Left. A number of respondents mentioned possible reactions on the right, and several were concerned with the possibility that "immediate withdrawal might lead to making the American military look like fools."

The most important domestic consequence of the Vietnam situation, in the eyes of some respondents was

that as a result of the war in Vietnam, the war on poverty, racism, and social priorities were [sic] being slighted. I concluded in 1967 that as long as the war in Vietnam went on, it was impossible to do something about these other ones. The war also diverted political and what I would call moral energy.

Thus far we have seen the pragmatic argument to encompass the following points:

1. We cannot win in Vietnam because American power is limited (expressed by 46 respondents).
2. Because American power is limited, to waste this power when it cannot prevail endangers the free world (expressed by 5).
3. Prolongation of the war splits the country with both right and left opposing American policies, thus further weakening the United States both objectively and in the eyes of other nations, including the Communist countries (expressed by 17).
4. Concentrating American energies on Vietnam results in an inability to work out our domestic problems (expressed by 6).

Pragmatists do have a political position, one which is essentially liberal rather than radical, and which is not opposed to some kinds of intervention.

There is one pragmatic position, frequently mentioned along with one or another of the specific pragmatic objections to the war, which neatly sums up all these objections to our policy as well as pointing up the liberal orientation to the United States. This summary position can be called the doctrine of American interest, and was expressed by 12 respondents. In this view, which comes close to being an ideology rather than simply pragmatism, the United States had no interests in Vietnam and in fact its presence there was contrary to its national interests. Frequently this position implied a strategy for dealing with the problem or was in overt opposition to moral or ideological bases for opposing the war. The impassioned exposition which follows explains the position far more ably than we can, and shows how pragmatism can become almost a formal ideology.

> I don't think it is in the self-interest of the United States to be there; I'm not being moral, there is a point where I would get moral about the morality of our position, but I don't think it is in our self-interest. . . . We are being cleverly sucked in and exploited by certain factions of the Chinese and Soviet governments who have an interest in seeing us keep off balance. . . . The only thing we can do to *save our wealth and power* is to pull out.
>
> *This pullout can only be accomplished by convincing the majority of people that it is the best practical solution. . . . You are not going to convince them to pull out by talking to them about the war the way Chomsky does.* . . . But in any case, his arguments are altogether too rational and undifferentiated and theoretical to convince not only the people, but me or anybody that runs the government that this is a proper reason for pulling out. . . .
>
> *I don't think the change of consciousness that people want to bring about in this country is going to be brought about by our being defeated in Vietnam.* . . . I think you start by making things appear . . . profitable. [Italics supplied]

Now we come to more formal ideological grounds for opposing the war. All told, 37 respondents had specific, overt ideological objections to the war, though, as we saw, most of the pragmatists also had a more hidden ideological motivation. Radicals, as one would expect from the very definition of radicalism, generally opposed the war on ideological grounds: that is, they questioned the basic premise of the policy. When asked whether he thought sending American troops to Vietnam was a mistake, a "libertarian Socialist" replied:

> I was opposed to the war from the beginning on much more general grounds and consequently I don't think it's meaningful to ask me whether I think the policy was a mistake. . . . I can see where many of them have come to the conclusion that . . . it was a blunder. But that's their problem, not mine.

This type of ideological reply specifically rejects any pragmatic framework. Another respondent, once active in the labor movement, said, "It strikes me that critics of American foreign policy in Vietnam are absolutely justified in characterizing this as an extension of nineteenth-century imperialism, which is mindless." Only six respondents gave such leftist reasons for opposing the war.

Pacifism is another kind of anti-interventionist ideology, which, though based on a moral principle, is generally elaborated into a philosophic system. There are relatively few in our sample (4), though they all tend to be quite articulate.

> I am an anarchist pacifist. I think people should be left alone. . . . I don't think any rights or wrongs in Vietnam were ever as bad, no matter how wrong the Communists were or anybody else, compared to the amount of evil which is done when some great power intervenes . . . So that even if our intentions had been the most sublimely beautiful in the world we are bound to do more damage than good.

Another type of political ideological objection to the war came from Democrats and democratic socialists. Rather than objecting to intervention per se, they opposed the aims of this *particular* intervention, although they, too, opposed "imperialism." This position was much more popular, with 16 holding to some version of it. One socialist explained that he became concerned about Vietnam when advisers were sent in. He thought to himself, "Here is another example of support for the right wing." The question of when he became concerned, however, was "difficult to answer since [he] had been active in the radical movement back to the Korean war" which he had also opposed for ideological reasons. An editor who was not a socialist explained his opposition: "I don't think it is a sufficiently democratic government that the majority of the people in South Vietnam want . . . they might prefer a Communist government. . . . We should accept rather than defend whatever they choose."

Important variations on the "imperialism"-and-intervention theme came from persons who, by no means radical or socialist, also came to oppose the Cold War. Some found it passé, and therefore not a very good reason for policy-making. "This is a very subtle question, but very broadly, I think the issue of Communism as such, as a prime symbol of evil, has very substantially subsided. . . . Even Nixon doesn't talk very much about Communism, not as much as Johnson does. He talks about American honor and not accepting defeat. I think this is an important point." Others, more fervent in their opposition, were against the war in Vietnam because they came to believe that the Cold War was never tenable.

The last ideological position which we encountered among our respondents was expressed by only two persons and represents what one might call neo-isolationist. This position in principle opposes any relation between the United States and Asia or perhaps the entire world.

Despite the moral fervor of American college students, only a small minority (one fifth) of leading intellectuals objected to the war on moral grounds. Interestingly, intellectuals who morally object to *all* wars or to *all* killing tended to offer an ideology or a philosophic system for this stance. That is, war or killing is wrong because they are an instance of something else; they are not immediately wrong absolutely and in and of themselves. The only direct, purely moral objections to the war in Vietnam were objections to *this* war or to some aspect of it. Respondents, themselves, recognized the difference between abstract philosophical-ideological objections to the war and direct moral objections. The respondent who was quoted earlier as a pacifist and who offered objections to war in general also had some specific moral comments to make about this war in particular. They are in a different vein from his anarchist-pacifist ideology, and he recognized the difference himself.

> With regard to the Vietnam war . . . there is so much in it that is *not political but actually moral,* that is, it rouses indignation . . . dropping napalm on simple folk and drafting kids to get killed there, etc. There are moral issues and, of course, one has to join organizations to stop an immoral war. I have been, as you know, with organizations that have urged active resistance to the war. . . . But that doesn't seem to me as a method so much of changing the policy but as of just acting in a moral way. . . . When an outrage is occurring you have to do something. [Italics supplied]

Morality brings with it an urge to immediate action, even if the action is a relatively mild one, as in the case of the aging poet who did probably all he was capable of doing:

> We shouldn't have been in there in the first place, even if we'd have succeeded. I think it was a criminal thing to do. Particularly with the misery, the wretchedness that we have contributed. I might read to you a little bit of one of my poems on the subject, a few lines, maybe.

The theme of the irrelevance of our ability to win in Vietnam was often repeated by those offering moral objections. The implications of American firepower were especially distressing to some. An art critic said:

> They were dumping bombs on people just to get rid of them. And then they also developed this strange criminal psychology of killing people for the fun of it.

A professor of economics summed up his reasons for changing from a supporter of the war to a person strongly opposed:

> I guess it must have been the reading [about Vietnam] and a growing sense of moral outrage about the disparity of firepower. You know, this American Colossus on the peasants. I'm sure that is a very strong factor, other than the politics of the situation.

In sum, almost 80 percent of the 99 intellectuals who gave us a full set of reasons for opposing the war opposed it on pragmatic grounds, almost 40 percent opposed it for ideological reasons, and a bit more than 20 percent for moral reasons. Obviously, this adds up to more than 100 percent, because some intellectuals offered several reasons. Indeed, almost everyone offered pragmatic reasons, so the distinguishing characteristic is whether or not some other kinds of reason were also advanced.

Let us call an intellectual a "moralist" if he gave a moral reason, even though he may have offered other ones as well; now, peeling off the layers of the onion, the next layer consists of "ideologists" – those 30 percent of the sample who did not give a moral reason but who did give an ideological reason for opposing the war; finally a bit less than half gave *only* pragmatic reasons and so are called "pragmatists."

We shall offer one important observation on this categorization of reasons for opposing the war in Vietnam which will not be repeated but which is absolutely crucial to understanding our entire analysis. We are dealing with reasons *volunteered* by our respondents in the course of their discussion of the Vietnam issue. We did not specifically query intellectuals and ask them, "Do you take a moral, ideological, or pragmatic view of the war?" When pressed, most intellectuals might admit that the war is immoral, or even that it is ideologically wrong. The point is not what reasons they gave when pressed, but their spontaneous reactions, their natural way of thinking about the war.

The reasons given for opposing the war in Vietnam intertwine with every aspect of an intellectual's orientation to the war. Vietnam historiography is strongly related to reasons for opposing the war. Intellectuals who were opposed to the Cold War in the 1950's and who felt that Cold War ideology was the cause of our intervention in Vietnam quite naturally themselves opposed the war in Vietnam for ideological and moral reasons. On the other hand, almost all of those who did not oppose the Cold War in the 1950's but who nonetheless thought that the Cold War was responsible for our entry into Vietnam opposed the war only for pragmatic reasons. That is, what was wrong with the Cold War as extended to Vietnam was that it did not work, not that it was ideologically or morally wrong. And certainly, those who felt we entered Vietnam as a result of step-by-step miscalculation also opposed the war because we could not win it.

The stance intellectuals took on the Cold War in the 1950's also affects their *reasons* for opposing the war in Vietnam at the time of the interview. Only ten percent of those who opposed the Cold War in the 1950's opposed the war in Vietnam for pragmatic reasons; most opposed it for ideological reasons. The converse is true of those who had been Cold War "Stalwarts" in the 1950's (positions one and two). In 1970, 70 percent opposed the war in Vietnam for pragmatic reasons. Those in between on the Cold War remained in between: they were likely to have offered every reason for having been against the war in Vietnam.

The flexibility allowed by a pragmatic orientation was nowhere more evident than in considering who changed his mind when about the war. Although some of the middle switchers — those who changed their minds about the war between 1963 and 1965 — offered moral reasons for their change, almost all the switchers and almost all those who still favored United States policies in Southeast Asia were pragmatists. In contrast, almost all of those who were always opposed to the war from its very start opposed it on ideological grounds, for ideology always allows one to have a firm, immediate, and relatively fixed opinion. Those who also consistently opposed the war but did not become attuned to it until somewhat later — again between 1963 and 1965 — were more likely to oppose the war on moral grounds. A moral reaction is not only generally slower in developing but in this case, the moral evidence on the war was not as easily or as widely available in the early period.

With this basic review of reasoning on the war, we can return to our initial concern with the policies intellectuals proposed for Vietnam. American intellectuals with different positions on the war indeed had quite different reasons for taking them. Two thirds of those who wanted to get out immediately objected to the war on either ideological or moral grounds, with ideology predominating. The bare majority of those who would have negotiated and worked for a coalition government would have done so on pragmatic grounds, with the remainder split between ideology and moralism. Almost all those who would have liked to prevent a Communist takeover talked about the war in pragmatic terms.

Do pragmatists, moralists or ideologists have some special characteristics that make them more likely to hold to these orientations? Generally, no. To be sure, all the members of the radical circle took an ideological position, all the members of the small *Partisan Review* circle in our sample took a pragmatic position, and none of these, as well as none who are in the influential center circle took a moral position. Otherwise, pragmatism, ideology, and moralism were all more or less spread through the literary and the literary-social science circles, as well as among non-circle members. Nor was there any appreciable relation between social background and style of argument on Vietnam. All of this further suggests that the war in Vietnam has effectively shaken up the traditional patterns of American intellectual life.

One thing about intellectual life and Vietnam is clear, however. The leading American intellectuals are pragmatists, *not* moralists. Not only did pragmatism dominate recent history – it was the ruling mode of thought in the fifties – and not only was it the most commonly held position in 1970, but pragmatism also dominated the thinking of those intellectuals who most influenced their fellow intellectuals on the war in Vietnam. Over 75 percent of those 17 intellectuals in our sample who received two or more "votes" from their colleagues as having influenced them in some way on the war were pragmatists. Even more striking, moralists simply did not count. Not a single one of the 17 Vietnam influentials advanced any moral reason in our interviews for opposing the war in Vietnam. Those who gave moral arguments either did not at all participate in the discussion of the issue or were on the periphery. Moralizing is not the sort of argument intellectuals feel is convincing.

Despite the general triumph of pragmatism, there are some signs in the American intellectual community of a move toward ideology, and in this respect the core group in the Vietnam discussion may represent a future direction. Though 60 percent were pragmatists, 40 percent were ideologists; there was one lonely moralist in the crowd.

Since our analysis of the ideological content of the core discussion group on Vietnam was made, several of the members have publicly, in print, changed their views toward a more moral direction and some have tied critical comments on Vietnam and Southeast Asia to fundamental critiques of the American power structure. The significance of these changes and the general impact of the most influential intellectuals on Vietnam will be discussed in the next chapter.

7

The Vietnam Influentials

The culture of business has good reason to make use of the aphorism "It isn't what you know; it's whom you know." The culture of intellectuals might well offer its own version: "It isn't what you say; it's who says it."

Judging the worth of ideas is a great effort, and one can save much energy by paying attention to the ideas of those persons whom one has previously noted as having ideas worth listening to. The reader has heretofore been deprived of his chance to judge for himself the worth of the people who have advanced various ideas on Vietnam because we cannot release the names of our respondents. However, we can, to some extent, redress the injury caused by the anonymity of our respondents by talking, instead, about the persons they named as having influenced them. Some of the influentials were also respondents, but because we followed a sampling procedure, many of the influentials were never even approached by our staff. We shall be talking about a set of leading intellectuals, not necessarily respondents in our study, who have published a great deal on Vietnam, who enjoyed some measure of public fame, and whose statements can *directly* be attributed to them.

Our interest is in ideas, not gossip. Since journals were a major source of new ideas on Vietnam for the American intellectual elite, especially in the early period, we can begin to zero in on the content of the ideas of the leading intellectuals by noting which journals were named in response to our question, "Which three journals have had a special impact on thinking about Vietnam?" The rank order of the journals is given in Table 17.

There are some surprises in view of the list of journals most often read by our sample (see Table 17). The *New Yorker* rates higher than its overall position as an intellectual journal, the *New York Times Magazine* and the *New Republic* somewhat lower, and *Harper's* markedly lower; the *New York Times Book Review,* generally read, was simply not seen as a source of influence on Vietnam. Except for *New York Review* and *I. F. Stone's Bi-Weekly* – now appropriately merged – none of the journals on this list has been especially radical on the subject of Vietnam. Most have stressed the stupidity of American policy

TABLE 17

Journals Most Often Mentioned as Influencing Thinking on Vietnam (83 Respondents)

RANK	JOURNAL	PERCENT MENTIONING JOURNAL
1.	*New Yorker*	39
2.	*New York Review of Books*	37
3.	*New York Times Magazine*	28
4.	*New Republic*	24
	I. F. Stone's Bi-Weekly	24
5.	*Commentary*	17
6.	*Atlantic*	14
7.	*Harper's*	13
	Dissent	13
8.	*Foreign Affairs*	12
9.	*New York Times Book Review*	11

without necessarily offering a basic attack on American institutions.

Despite the fact that books changed the minds of almost none of our respondents, a list of the books they had read is extremely important to an understanding of their thinking. The list given in Table 18 clearly shows why.

Except for Fall's *Two Vietnams,* the most widely read book, all the books had appeared *after* the majority of our respondents had already made up their minds. But, as the table also shows, almost all the books had already appeared in somewhat different form in one or more journals. It was the earlier appearance which was crucial. In fact, many respondents first said that they had read no books on Vietnam, then realized that in fact they had already read many of them when they had first appeared as journal articles. The rank order of the books and the journals fairly closely parallel each other. If one counts Fall as a *New Republic* writer, then the top four journals are the sources for the books of the top four writers. Morgenthau, of course, had switched to the *New York Review of Books* by the time of our study, but had we surveyed our respondents earlier, then *Commentary* might well have come up higher in the ratings, because by the time of our study it had become "bored" with Vietnam. Also important to note is that articles by the Vietnam and foreign policy experts – Fall, Halberstam, Morgenthau and Shaplen – all appeared at least two years before the more generally known intellectuals, such as McCarthy, Draper, Schlesinger, Sontag and Galbraith, made their appearances.

The important thing about these ratings is that they allow us to compare the ideas of the most influential intellectuals on Vietnam with those of our sample generally, and they allow the reader to check up on our statements, and disagree with us, if he reads the books in a different way. Books tend to reprint in revised form the most important journal articles, and, on second thought, their more permanent nature suggests that they reflect more of what an author really believes. By analyzing the leading books it is fairly easy to see what ideas in print American intellectuals were exposed to. A quick review of the names of the authors immediately suggests that the general

TABLE 18

Books on Vietnam Read by 96 Leading American Intellectuals by 1970 in Order of Frequency Mentioned

RANK	AUTHOR	TITLE (WHERE ORIGINALLY PUBLISHED)	YEAR OF PUBLICATION	PERCENTAGE READING
1.	Fall	*The Two Vietnams*	1964	55
2.	McCarthy, Mary	*Vietnam* (*New York Review of Books*)	1967	44
3.	Halberstam	*The Making of a Quagmire* (*New York Times*)	1965	41
4.	Morgenthau	*Vietnam and the United States* (*Commentary*, others)	1965	36
	Shaplen	*The Lost Revolution* (*New Yorker*)	1965	36
5.	Draper	*The Abuse of Power* (*Commentary*)	1967	32
6.	Lacouture	*Vietnam: Between Two Truces* (*Le Monde*)	1966	28
	Schlesinger	*The Bitter Heritage*	1967	28
	Sontag	*Trip to Hanoi* (*Esquire*)	1969	28
7.	Fulbright	*The Arrogance of Power*	1966	26
8.	Goodwin	*Triumph or Tragedy* (*New Yorker*)	1966	25
	Hoopes	*The Limits of Intervention*	1969	25
9.	Galbraith	*How to Get Out of Vietnam* (*New York Times Magazine*)	1967	24
	Schell	*The Village of Ben-Suc* (*New Yorker*)	1967	24
10.	Pike	*Viet Cong*	1966	21
	Salisbury	*Behind the Lines* (*New York Times*)	1967	21

drift was liberal, rather than radical. (Chomsky's *New Mandarins,* though published in 1969, had not yet entered the intellectual mill as a book at the time of our interviews, nor had I. F. Stone written a book on Vietnam, a fact which somewhat, but not crucially, as we shall see, biases this following account.)

To verify this impression, we took all sixteen books in the top ten ranks (in Table 18) in our sample and checked out, wherever possible, the main drift of each book on the three key matters for which we have interview data: why the United States entered Vietnam, why the war should be opposed (or supported), and what current United States policy should be. Table 19 shows the tabulation of ideas about the war of the leading books, as compared to the issues of the leading intellectuals in our sample.[1]

TABLE 19

Opinions on Vietnam War in 16 Leading Books Read by American Intellectuals Compared with the Opinions of the 110 Intellectuals

	LEADING BOOKS	ELITE INTELLECTUALS
Reasons for Entering		
Cold War	6	24
Step by step, error	6	12
Individual decision-makers	2	10
No answer	5	64
Reasons for Opposing or Supporting		
Pragmatic	10	78
Ideological	4	38
Moral	2	22
No answer	1	11
Future United States Policy on Vietnam War		
Immediate withdrawal	3	45
Negotiate, deescalate	11	50
Continue, prevent Communist takeover	1	14
No answer	1	1

Book authors were less likely than the rest of the leading intellectuals to blame Cold War policies and ideologies for entrance into Vietnam and more likely to favor the more conservative step-by-step, blunder-by-blunder explanation. About the same proportion of authors and leading intellectuals offered pragmatic, ideological and moral reasons for opposing the war, but the authors of the leading books were more likely than leading intellectuals to favor negotiation rather than simple, immediate withdrawal. This difference is probably caused by the greater optimism about negotiation which prevailed at the time most of the books were written. If we extrapolate from the opinions of the authors in our sample given to us in 1970, then the book authors as a group do not differ from the rest of the leading intellectuals. In any case, it is safe to say that the climate of opinion engendered by the leading books on Vietnam is surely not overwhelmingly radical – in the sense of mounting a basic ideological or moral attack on American institutions – and certainly no more radical than that of the sample of leading intellectuals we interviewed.

Not every intellectual whose opinions on Vietnam were influential wrote a book, of course. Some wrote mainly articles, some wrote books but were chiefly thought of as article writers, and some were active in personal relations or in politics. The list of intellectuals influential on the Vietnam issue therefore needs to be expanded to include those whose influence was exerted through many channels. In the list of influential intellectuals which follows we counted every person mentioned in our discussions about the respondent's initial and subsequent reactions to the war, including many face-to-face contacts. This list does *not* include persons who were mentioned *only* as discussion partners, since to do so, we feel, violates the confidences given us. Naturally, each influence was counted only once. Thus, if a respondent replied, "Bernard Fall," in answer to every question about influence, Fall would still be counted only once in the table which follows. The number opposite each influential intellectual's name, then, represents the number of *different* respondents who mentioned him.

As a further aid to interpreting the table, you must know that not every influential was mentioned because a respondent necessarily agreed with him or her. Recall the negative remarks about Chomsky, McCarthy, and Sontag made by a respondent quoted above (Page 139). Others specifically mentioned Joseph Alsop as a person they follow closely in order to think the exact opposite of what he says! Other persons were also mentioned because they influenced the respondent in ways different from what was intended. For example:

> The second book [which made me rethink my ambivalence about immediate and direct withdrawal] is ——. I found it a terribly rotten book, but because he happens to be sort of stupid personally, he gave an extremely sympathetic view of the Viet Cong, although I bet he didn't realize that. In fact, it made me reinterpret my own view of it [the Viet Cong].

Others, of course, mentioned the very same writer because they thought highly of him in a more conventional fashion. In any case, it is obvious that even those with whom a respondent does not agree can be highly influential in his thinking.

Finally, just because someone is on this list does not mean that he has had an earthshaking influence upon policy. While the issue of the relations between intellectuals and political policy-making will be discussed below in Chapter 14, the following view of a university professor who was involved in the Kennedy administration is worth remembering:

> I don't think intellectuals have a great deal of effect. I mean, take the people in the Johnson administration, the people Tim Hoopes describes in his book. . . . I think that Hoopes and Nitze and people like that changed their views on Vietnam not because of anything that any intellectual wrote, but because of the evidence of the failures of the policies which they saw every day. . . . I think it is a mistake to assume a bilateral relationship between intellectuals and opinions. It is a triangular relationship between events, writers, and opinions. And events are the most important. The Vietcong have changed our minds about Vietnam, not the people that have written. I think that Robert Kennedy and Eugene McCarthy, McGovern, etc., had more effect in changing views about Vietnam than Galbraith, Goodwin and Schlesinger.

Now, to the list itself. We could, of course, correlate material from our interviews with the rank of the influential, but to do so we would have to reveal which of these persons we actually interviewed — a revelation which we feel violates confidence. The list itself does not, however, violate confidence at all, since the characteristics of the sample are known only in a general way and a given respondent cannot be linked to the naming of any given influence.

Leading Influencers of American Intellectuals on the War in Southeast Asia, 1970[a]

RANK	NAME	"VOTES"
1	Bernard Fall	56
2	David Halberstam	40
3	Robert Shaplen	36
4	Theodore Draper	29
5	Mary McCarthy	27
6	Hans Morgenthau	24
7	Jean Lacouture	21
8	Arthur Schlesinger, Jr.	18
	Jonathan Schell	18
9	I. F. Stone	17
10	Harrison Salisbury	14
	Richard Goodwin	14
11	Douglas Pike	13
12	Susan Sontag	12
	Walter Lippmann	12
	Richard Rovere	12
	Noam Chomsky	12
13	Edwin Reischauer	11
14	Joseph Buttinger	10
	Townsend Hoopes	10
	George Kahin	10
	Howard Zinn	10
15	John Kenneth Galbraith	9
	Walt W. Rostow	9
	James Reston	9
16	David Schoenbrun	8

RANK	NAME	"VOTES"
17	Malcolm Browne	7
	Marcus Raskin	7
	Tom Wicker	7
18	McGeorge Bundy	6
	Frances FitzGerald	6
	Irving Howe	6
	John Lewis	6
	Franz Schurmann	6
19	Joseph Alsop	5
	David Brinkley	5
	Chet Huntley	5
	Robert Scheer	5

[a] All intellectuals mentioned five or more times in connection with the following questions:
22. When did you first become especially concerned about this issue?
23. What were your views on America's involvement at this time?
24. What events, personal experiences, books, journal or magazine articles, newspaper stories or editorials, individuals, groups or organizations made you especially concerned about America's involvement at this time?
27. Since your initial concern, what three books that you have read had a special impact on your thinking about this issue?
31. Are there any particular authors or articles in these journals that stand out in your memory as especially important to your thinking on Vietnam?
32. Since your initial concern, are there any particular new media correspondents or columnists who stand out in your memory as especially important to your thinking on Vietnam?
38. Is there anyone else who influences your thinking?
There were 110 respondents, of whom 109 mentioned someone. No matter how many times a given respondent mentioned the same person, in the above tabulations he appears *only once* for each respondent.

Without giving away anything that should not be given away, it is possible to compare the characteristics of the intellectuals whose opinions on Vietnam were influential whom we sampled, or whom we actually interviewed, with the rest of the intellectuals in our study.[2] As compared with the rest of the elite intellectuals, the Vietnam influentials are clearly experts, rather than generalists. Although there is some relationship between presence on this list and membership in the top 70 (above, Pages 30–31),[3] as a casual inspection might reveal, the present list clearly includes many who first made their reputations as Vietnam experts. Compared with other elite intellectuals, persons with two or more mentions as having influenced the opinion of our respondents are:

(1) more likely to have written or talked on Vietnam;
(2) more likely to have used special channels in an attempt to influence Vietnam policy;
(3) more likely to have been generally active politically;
(4) but only slightly more likely to have been active in organizations opposing the war.[4]

Further showing their expertise, those with more nominations were more likely to:

(1) have taken their Vietnam position as a result of firsthand experience or research;
(2) have scored high on our Vietnam information scale. But they were no more likely than the average elite intellectual to be passionate in their opinions on the war. As befits their position of leadership, they are likely to be much wealthier than their colleagues, with much of the difference due to royalties.

As for objective social position, the experts were barely different from the rest of the intellectuals. There is *no* generational difference. And, though Vietnam discussion partners are more likely to be New Yorkers and to be Jews (remember, the two are *not* related among leading intellectuals), influentials on Vietnam do not exhibit these traits. On the other hand, the influentials are slightly more likely to be nonacademic.

Much more important than their expert character or their lack of important social differences from the rest of the respondents, influentials on Vietnam on the whole simply *do not differ from other intellectuals in their views of the Vietnam War in any substantial way*. That is, they include about the same proportion of those who wanted to get out of Vietnam immediately, as compared with those who wished to negotiate. There were, however, fewer who wished to hold back the Communists. Nor is their style of argument on Vietnam significantly different from the rest of the intellectual elite. There was a slight, but not significant, tendency for the Vietnam influentials, themselves, to say they were more radical or more strongly liberal than other intellectuals. In this vein, there were fewer who held extreme conservative positions on the Cold War in the late fifties among the Vietnam influentials, but again, these are differences of no more than 10 or so percentage points. Further, when we scored them on the way they presented themselves ideologically on other issues, we found no differences between the influentials and the others. Nor do the influentials come more frequently from any one of the more permanent intellectual circles (though discussion partners were more likely to be from the center circle). In short, this analysis confirms our picture of the leading books: the intellectual elite in our sample share the views of the intellectuals they said had influenced them. To a student of the social psychology of leadership, these findings that leaders and followers in a system of high interaction have the same opinion can hardly come as a surprise. But they do reinforce our picture of elite intellectuals on the whole as not having taken radical positions on the war in Vietnam.

The terms "radical" and "nonradical" have been bandied about. An examination of the list of names and a check on what the influentials have written may be useful. Also useful in this context would be to compare the list of intellectuals who most influenced our respondents with a list the respondents thought most influenced the intellectual community on Vietnam. The first list is a list of *actual* influentials, the list in Table 20 is a *reputational* list. Comparisons between the lists suggest that in 1970 the American intellectual elite perceived radicals as being much more influential than they actually were.

Our definition of radicalism on foreign policy coincides with Robert Tucker's.

> The essence of the radical critique is not simply that America is aggressive and imperialistic, but that it is so out of an institutional necessity. It is the central assumption that American imperialism must ultimately be traced to the institutional structure of American capitalism that is the common denominator of radical criticism. It is the same assumption that most clearly separates radical criticism from all other criticism, whether liberal or moderate left.[5]

TABLE 20

Intellectuals Most Influential on the Discussion of the Vietnam Issue within the Intellectual Community, as Perceived by Peers

RANK	INTELLECTUAL	"VOTES"
1	Noam Chomsky	38
2	Hans Morgenthau	25
3	John Kenneth Galbraith	19
4	I. F. Stone	18
5	Mary McCarthy	16
6	Bernard Fall	15
	Arthur Schlesinger, Jr.	15
7	Theodore Draper	9
8	Jean Lacouture	8
	Norman Mailer	8
9	Richard Goodwin	6
	Dwight Macdonald	6
	George Wald	6
	Susan Sontag	6
10	Paul Goodman	5
	Staughton Lynd	5
	Edwin Reischauer	5
	Richard Rovere	5
	Harrison Salisbury	5

It is especially important to note the distinction between radical left and what Tucker calls "moderate left" or the group that has been identified over the years as democratic socialists. Tucker puts it this way in a footnote to the above:

> It hardly seems appropriate to label as "radical left" such writers as Michael Harrington, Arnold Kaufman, William Pfaff, Robert Heilbroner, and Irving Howe. . . . In their views of foreign policy, if there is anything that holds them together (with, perhaps, the exception of Harrington), it is the belief that the problems of American foreign policy cannot be expected to yield to some simple explanation

— and solution. And they cannot be expected to do so because they cannot be reduced to any one root or basic source which, once perceived, may then ultimately be transformed. . . . *Whereas the radical left finds American foreign policy irredeemable so long as American society remains essentially unchanged, the moderate left does not believe that significant change in foreign policy must await a basic transformation of American society.*[6]

Both the moderate left and the radical left tended to offer ideological objections to the war, along, of course, with pragmatic ones. But the content of the ideology and the degree of pragmatism differ considerably. When we talk about radicalization and radicals, we mean that the war in Vietnam was seen as a direct result of the basically wrong character of American institutions. The "wrongness" may be economic, it may be social, or cultural, it may even be anomic or alienated personalities, but this wrongness must affect, and be the cause of, the basic decision-making process of American foreign policy. Thus, we include not only orthodox Marxists within the "radical" definition but various styles of revisionist Marxist thinking and even some of the more amorphous "New Left" critique.

Few in the ranks of the top dozen most influential American intellectuals on the war in Vietnam approached the discussion of Vietnam from a radical position. Fall argued all along that Ho Chi Minh was Communist first and a nationalist only second; Halberstam in his collection of dispatches merely pointed out that the United States was not accomplishing its announced goals, that it was losing rather than winning; Shaplen thought the United States could and should stay in Vietnam and in other developing areas in order to encourage these nations to develop democracy;[7] Draper conceded that the NLF was Communist-dominated and that it had popular support in the South. The United States, according to Draper, simply could not hold onto every status quo situation in order to maintain the line against Communism, for to do so would be to squander its powers. Mary McCarthy argued from an essentially moral position that the United States was savagely destroying and corrupting South Vietnam's land, culture and society out of sheer ignorance and self-delusion, but noted "the uselessness of our free institutions . . . to interpose any check on a war of this character," and Morgenthau argued that we had neither national interest in Vietnam nor the ability to muster adequate power. Of the remaining names on the list of the top dozen, Schlesinger, Salisbury and Goodwin are well-known liberal Establishment figures. Schell offered a scathing attack on the morality of the United States effort in Vietnam. I. F. Stone, an independent leftist, is perhaps hardest to classify in "radical-nonradical" terms, since most of his writing on Vietnam and other foreign affairs has consisted of carefully documented muckraking, showing the deceits, as well as the devious intentions, of American leaders. Lacouture was the only non-American in the top dozen and so is least concerned about domestic American affairs. He did point out that the NLF originated out of Diem's repressions and that in order to gain any satisfactory solution, the United States would have to give up the idea of setting up an anti-Communist government in the south. Thus, if one stretches the term, one might count as radical in their critique of American policies on Vietnam only Mary McCarthy, Stone, and possibly Lacouture. None of the others, with the possible exception of Draper, basically attack the social fabric of the United States; rather, the common argument is *realpolitik:* the United States could not possibly attain its desired goals of containing Communism or at least of supporting democracy in the fashion it is pursuing in Vietnam. This ratio of one-to-four is even lower if one looks at all 38 on the list. Again, with a broad interpretation of the term, Sontag, Chomsky, Zinn, Fitz-Gerald and Scheer seem to have published radical critiques of the war, thus giving a total of eight radical critiques. If one wants to add a couple of more names, on the ground that we have not been generous enough in our definition of radicalism, then we have 10 of 38, or about one quarter. Compared with any other group in the population, a figure of 25 percent radicals is high, indeed. Compared with the intellectuals' own perceptions of themselves, this is low, indeed.

The major difference between the way intellectuals think of themselves and the reality reflected by their own reports is the replacement of Fall by Chomsky. Fall is clearly nonradical in his critique: Chomsky is clearly and self-consciously radical. But Chomsky ranked only in the twelfth position in terms of actual influence. In terms of *perception* of influence he is without peers. Further, the proportion of intellectuals who offered radical critiques of the war is somewhat greater on the list of perceived influentials. We count 6 of 19, or somewhat less than one third, of those receiving five or more "votes," as compared with between one fifth and one quarter of those receiving more than 5 mentions as having actually influenced our respondents. Aside from the undoubted perceived prominence of the *New York Review of Books* and its association with Chomsky, which has boosted his prominence, the list of *perceived* influentials is much more affected by the *general* prestige of an intellectual. Cases in point are Dwight Macdonald and Norman Mailer, neither of whom appears on the list of actual influentials, but both of whom are prominent intellectuals who were very visible on the picket lines, though not, in our view, radical critics of the war. Both were very vehement in their published statements on the war and wanted an immediate "out" as early as 1965, but neither, in their hatred of Communism's version of dictatorship, was willing to castigate the American system and to praise Ho Chi Minh, as were some (but not all) of the radical critics.

If one asks intellectuals about the sixties, most would agree with this old leftist who felt that his faith (or, more exactly, his lack of faith) was renewed. "I think that the Vietnam War, together with certain other events, served to radicalize me, along with a number of other people who, perhaps, in 1948 and 1947 had accepted the notion that American society was improving." Thus their own perceptions were that the entire intellectual community had moved to the left. That the community had become much more willing – nay, even forced – to criticize American policies there is no question. But dissent does *not* equal radicalism. Those who most effected the change were not the new radicals, the new activists, or the newly-revived activists, or the revisionist historians. Rather, the word came most of all from within the liberal group of the leading intellectuals, themselves. And, for the most part, those who were among the

Left in the group of leading influentials on Vietnam came not from the New Left, but from the old democratic socialist Left, the very group the New Left had been so assiduously attacking.

Perhaps the nonradical caste of the list of influentials is best explained by the following quotation from an interview with a social scientist who was an early opponent of the war in Vietnam, on pragmatic grounds, but who had basically supported the Cold War in the 1950's.

> I think I might have read Susan Sontag's "Trip to Hanoi" in *Esquire*. . . . I read Staughton Lynd and Tom Hayden in *Viet Report*, but, you know, the funny thing is that I wouldn't believe the people with whom I disagreed politically.

To have credence, an anti-Vietnam War position had to come from the ranks of those who were politically "respectable." The mass media and some of the leading intellectual journals may have given special play to radicals, but the majority of intellectuals could trust only those who had won their spurs in the 1940's and 1950's.

So much for radicalism on Vietnam in the sixties. The war went on into the seventies. Did the continuation of the war lead to an increase in ideological and moral thinking by leading intellectuals? Did the apparent success of Nixon in withdrawing, and eventually securing peace, take some of the wind out of the advocates of the "we-can't-win" position? Our impression from published sources is that there has been little change. The *New York Review of Books* has continued to publish ideological and moral criticism of the war, with a most notable cry of outrage coming from an old-timer who had not been previously prominent in the debate – Henry Steele Commager (October 5, 1972). Both Barnett and FitzGerald would now probably rate higher in actual influence, and both have criticized American policies from a radical point of view. But the views of those most representative of our sample have not appreciably changed. Fall, of course, was killed in Vietnam even before our interviews in 1970. Halberstam in his new book, *The Best and the Brightest* (New York: Random House, 1972), continues his chronicles of American blindness and stupidity, this time focusing on Washington, rather than on Vietnam. There is an Eastern American Establishment dominated by banking concerns, and this establishment is culpable. On the other hand, Halberstam never directly questions, in this book at least, the basic structure of the United States or what should be done about this establishment. By focusing on the politics of the bureaucracy, "Halberstam himself," as Victor Navasky noted in a sympathetic review in the *New York Times Book Review*,[8] "never defines the real center. . . . This is not a book about the destructive impact of the war on America; it is not a book about the immorality of the war."[9]

Perhaps the best sign of a basic lack of change in positions since our interview period is the Chomsky-Morgenthau debate reported in the Summer 1972 issue of *Partisan Review*. In his September 24, 1970, article in the *New York Review of Books* titled "Reflections on the End of the Republic," Morgenthau appeared to have shifted to a radical criticism of the United States – "The great issues of our day . . . are not susceptible to rational solutions within the existing system of [American] power relations." But by 1972 he reiterated his belief in the "possibility by rational analysis to arrive at certain objective conclusions which define . . . the national interest. . . . The policies we have been following in Vietnam are utterly destructive of the national interest." In short, a reiteration of the doctrine of national interest which we cited as the sine qua non of a pragmatic approach. Chomsky, on his part, reiterated that "what is called the national interest . . . is the specific, parochial interest of a particular ruling group. . . . It is by no means obvious that Kennedy and later Johnson intellectuals made the wrong decision from the point of view of the rational calculations of the American ruling groups. They provided the shield behind which this great miracle of capitalist economic progress in East Asia could take place."

Chomsky and Morgenthau agreed that they both militantly supported the same policies in Southeast Asia. Our argument, however, is that the ultimate effect of militant dissent from American policies and even militant advocacy of certain counterpolicies depends very much on the *reasons* for the dissent, and the *reasons* offered for counterpolicies. In the next chapter we shall show that the pragmatic orientation of American intellectuals profoundly affected the kinds of lesson they learned from the Vietnam experience.

8

The Lessons of Vietnam

As a result of the Vietnam War the American intellectual elite know what they do *not* want in foreign policy; they are much less certain or clear about what they *do* want. For the important thing about the dominance of pragmatic thinking among American intellectual leaders is that it allowed old Cold Warriors to have opposed the war in Southeast Asia with considerable vigor and even occasional militancy and yet continue to hold views of the world that are not appreciably different from those they held in 1960. We shall show that this basic continuity affects the intellectuals' reinterpretation of the past as well as the way they look to the future. Almost every one of our sample who talked about the Korean War supported it, and, contrary to the impression of the intellectual community one might gain through the reading of revisionist historians, almost no one in our sample has changed his mind about the necessity of the war and the basically correct stance of Harry S. Truman. The step-by-step stupidity of policymakers' explanation for the Vietnam War, together with pragmatic objections to its course and outcome, are essentially conservative positions. We shall see that the major lessons learned lie in the direction of caution, rather than boldness. At most, some suggested a major overhaul in the decision-making machinery. Few, indeed, suggested major changes in policy or in American social structure.

Southeast Asia remained an enigma to most intellectuals. Even the relatively few remaining ardent Cold Warriors did not uniformly endorse a policy of military intervention in Southeast Asia or anywhere else. Except for recent converts to an anti-Cold War policy who tended toward an isolationist posture, many were willing to consider some continued American presence short of direct military aid. The "Doctrine of Vietnamese Exceptionalism" seemed to be held by more members of the American intellectual elite than cared to admit it. As Robert Tucker pointed out in a review of published opinion:

> It is by now commonplace that the war in Vietnam led to a breakdown in the foreign policy consensus of the past generation. But commonplaces can be misleading. The lasting effects of the debate occasioned by Vietnam remain far from apparent.[1]

Our data, which include the opinions of many important intellectuals who have not published on foreign policy issues, reinforce Tucker's view.

Perhaps the most salient characteristics are uncertainty and confusion. Our analysis of the opinions of elite intellectuals inevitably lends an air of certainty to a situation that was far from clear to most of the persons we interviewed. The following from one of our transcripts is not atypical.

> [Q.] In view of the developments since we entered the fighting in Vietnam, do you think the U.S. made a mistake sending troops to fight there?[2]
>
> [A.] Yes, it was a mistake, in hindsight. I didn't think that at the time, let me tell you.
>
> [Q.] Would you like to elaborate on your position on America's involvement in the war in Vietnam?
>
> [A.] I think we made a disastrous miscalculation. . . . I changed my view. I'd argued the other side before that. I defended the war on the

basis that we had made — the national honor — a commitment, rather than the Communist menace in Asia.

[Q.] What were your views on America's role in the Cold War before its involvement in Vietnam?

[A.] I thought we were too unresourceful. I thought that we were always making a black-and-white confrontation where that wasn't necessary. Not just Dulles. . . . The Kennedy foreign policy fell short, too, in that it didn't have enough "give" in it. It was too much "us or them."

[Q.] Has the Vietnam war in any way changed your views on America's role in the Cold War?

[A.] No, I don't think so. I don't think that it has had all that much effect on my views of the Cold War.

Each respondent was asked for his prescription for Southeast Asia. The respondent just quoted favored a coalition government in Vietnam and was asked a series of probing questions. The interviewer's last try was, "You would say that America's policy, then, would be to encourage a new government other than the one that is already in South Vietnam? And if that is so, in what way?" He replied, "I don't know. You've gotten me into things now that I don't know enough about to answer. Step by step now you've got me out on this limb. I do feel that our support of the Thieu government is a mistake." This comment is a typical instance of knowing what you do not like, but not being certain of what you do like. We let most respondents talk in generalities about future policies because they found it difficult to go into any detail.

We emphasize the uncertainty of response because our classification of opinions about the future of the Cold War and American foreign policy may give a more coherent picture of respondent opinion than actually existed.

Korea

One of the most important lessons one learns from history is to change history itself. In the course of discussing the Cold War, almost half our respondents (47) talked about the Korean

War and their present reactions to it. That many more did not mention Korea, however, is rather surprising. The analogy between Korea and Vietnam was widely made by the administration in the early days of the war and one would imagine that many more might have defended or attacked this analogy. Yet almost all of the intellectuals did support American intervention in Korea at the time when it was begun in 1950 (only five were against it), although this intervention is now an embarrassment to many who have since taken a liberal position on the war in Vietnam. So that in many interviews, just as one thought the matter of Korea was about to come up, the respondent deftly shifted the topic or spoke in broad generality about the problems of Communist expansionism in the past without directly mentioning Korea.

One way of avoiding the issue was to plead ignorance. "At the time I had no real interest, I wasn't informed," or "I know nothing about Korea," or "I was a student then and wars were on the periphery of my attention." Although a considerable amount of revisionist history of the Korean War had appeared by the time of our interviews, not many respondents commented directly on it. This work, although read, had not yet become central to the thinking of most intellectuals; nevertheless, it had raised doubts. It remained in the footnote category, rather than as the central text:

> I must say, though, as a P.S. on this [Korea and the Cold War] that reading revisionist historians in the last couple of years really makes me wonder whether we went far enough in trying to understand the Soviet Union in that period.

We said that only five of the intellectuals who discussed it opposed the Korean War outright, at the time. Naturally, in retrospect, in 1970 they still opposed it. For the rest of the intellectuals who discussed Korea, the real question is how far were they willing to go in rewriting history. And the answer is, not very far at all. Only one respondent admitted to have wrongly supported our actions at the time. Although he was the only respondent who fully admitted that he made a mistake, he is the editor of an important journal, so that his personal opinion may carry rather more weight than that of a private person.

At the time I thought that we should [intervene] – when I thought of it as a U.N. police action. . . . I no longer see it in the same way. I feel disillusioned. . . . I thought actually, at the time, the U.N. was acting to prevent aggression and therefore to preserve peace everywhere by taking the action in Korea. And now, it doesn't look that way to me. . . . Because of our strong position in the U.N. we were able to drag others in along with us. . . . I've been much closer to being an absolute pacifist in the intervening years.

His changed opinion, then, was not caused by a reevaluation of the United States from a Marxian point of view as an imperialist power, but rather from a growing pacifist conviction.

There were another six converts to at least a position of ambivalence about Korea. A typical discussion is the following:

I was not utterly opposed to the Korean War when it happened. I'm not sure now that I was right . . . so long as the international aspect, the involvement of the U.N. and so on was real, it was a tolerable policy . . . certainly in retrospect I would think Acheson was wrong originally in proposing that the possible conquest of South Korea by North Korea would have been a desperate threat to the U.S. I do not think [now] that was so.

It is obvious, of course, that our respondent meant by "I was not utterly opposed" that he truly supported the Korean war! Another convert to ambivalence, if one can be a convert to a confused position, recaptured the strong emotional support given to the American GI at the time, support conspicuous by its lack during the Vietnam War.

I knew people who fought in that [Korean War] . . . and I . . . was proud of them for being good soldiers. . . . I'm sorry to say I have never explored the actual lowdown on the Korean War. I. F. Stone has a book which apparently makes our side of it look terribly shabby.

Some intellectuals claimed not to have been clear-cut in their support of the Korean War but they also were not clear-cut in rejecting it. An opponent of United States intervention in World War II and a staunch opponent of the war in Vietnam still felt confused about Korea.

> I don't think I was much for it. I wasn't against it, though. . . . It was different from the Vietnamese situation. . . . It wasn't a question of a mass popular movement (like the NLF) in South Korea at all – it was an invasion. And I felt you sort of had to stop it. . . . *It never became much of an issue.* [Our italics]
>
> [Q.] How do you feel about Korea today?
>
> [A.] I don't know. It's kinda hard, hindsight, you know. I really couldn't say.

The key, of course, is "It never became much of an issue." The sway of the pervasive disillusionment with the Soviet Union, the early years of the Cold War, and the reign of McCarthyism may have inhibited intellectuals from making an issue of Korea, even though both before and after Korea intellectuals tended to be vociferous opponents of United States policies.

Another form of ambivalence about Korea was to accept the American defense of South Korea but to oppose, even at the time, the extension of the campaign above the thirty-eighth parallel. The following comments by a leading opponent of the war in Vietnam who today is clearly a radical is typical of the five who took this position:

> Well, I thought you could give an argument for American involvement in the first three or four months until the thirty-eighth parallel was restored. . . . But after MacArthur reached the thirty-eighth parallel I was against it every step of the way.

Including the supporters of the thirty-eighth-parallel doctrine, about half of the intellectuals who discussed Korea had either always been opposed to it, or were at least ambivalent about it. Only one admits to have totally changed his mind and a few more changed from support to ambivalence – responsive to revisionist history.

What is the published point of view of a moderate revisionist historian?

The fact that the Korean conflict, initially, brought so many benefits to Truman, Chiang, and Rhee has led some historians and journalists [D. F. Flemming and I. F. Stone] to charge that Rhee started the war with covert support from America and Chiang. . . . These charges go too far. . . . The probability is that the Russians did not know the North Koreans were going to attack and in many ways the United States was better prepared to respond than the Russians were. What is true in the Flemming-Stone charges is that the United States knew the invasion was coming.[3]

At the United Nations, the Americans announced that their purpose was the simple one of restoring the 38th parallel. . . . The United States rammed its resolutions through the United Nations without discussion or investigation. . . . The United States established a military command that took orders from Washington, not the United Nations.[4]

This is obviously a different view from that given by respondents, and most experts who took revisionist views into account felt their arguments weak. Even those in our sample who were ambivalent about the war, or who are ambivalent in hindsight, emphasize the importance of the UN support for the action, support which revisionist historians, rightly or wrongly, claim was essentially phony. Almost all of the 22 intellectuals who clearly supported the war at the time or who clearly supported it both then and now, in retrospect emphasized the fact that South Korea was invaded and that it was a case of Russian expansionism – facts which revisionist history also strongly questions. Most supporters of Korean intervention also sharply distinguished it from Vietnam.

For example, "Korea was different. North Koreans were the aggressor – not like in Vietnam." And,

I was very much for the Korean intervention, but the situation was extremely different, as I view it. Among other things, Stalin was still alive and still in an adventurous kind of gambling mood and God knows what he would have pushed to if he didn't meet some kind of resistance there [in Korea].

Given these kinds of comment and the ambivalence of many respondents who are now strong opponents of the war in Vietnam, we must conclude that the large majority of those who did not talk about Korea must still basically support American intervention there, but find that support in some conflict with their present views on Vietnam. In short, whatever the war in Vietnam has done to change the opinions of leading American intellectuals about future foreign policy, it has hardly changed their views of the past.

Vietnam Lessons

If we are surprised that only half the intellectuals interviewed even mentioned Korea, we shall be even more surprised to learn that only half said they had learned anything from the Vietnam experience. There were those who said that Vietnam was a special case – one fifth; about two fifths learned that American power was limited; and another two fifths said they began to change their views on international Communism.

The doctrine of Vietnamese exceptionalism,[5] as it might be called, holds that something can, indeed, be learned from the war, but it consists of the fact that Vietnam, or more generally Asia, differs from Europe and that Cold War policies developed for Europe cannot or should not be applied to Asia. For example:

> [Vietnam hasn't changed my mind about the Cold War.] No. The Cold War concerned Europe, primarily. . . . Our ties to Europe were very much closer than to Southeast Asia. . . . It's very hard for Americans to get excited about what happens in Burma, Malaya, but we do have an understanding or feeling for what goes on in Germany, France or Britain.

A more fundamental reevaluation of the United States world position is the doctrine of "limitation of power" expressed by a foreign policy expert.

The Vietnamese war illustrates the extraordinary limitations of the use of power by a highly advanced society over a relatively backward, primitive society. This means that the whole concept of counterinsurgency, and so forth, has to be rethought. Second, it seems to me that the Vietnamese involvement indicates that the American people will not stand prolonged warfare, even of a limited type, and, hence, there will be overriding pressures, either toward military disengagement or rapid all-out victory – a kind of simplistic black-and-white approach to international politics which at times can be very dangerous.

If American power is limited, the lesson learners conclude, then there has been too much emphasis on the military.

The military were able to determine it [policy] because the politicians didn't know what to do. It is very difficult to formulate political policy for the U.S. in a part of the world which is not understood. . . .

The most radical consequence of the war in Vietnam was a growing disbelief in the doctrine of the "international Communist conspiracy." A reexamination of Cold War bellicosity often accompanies this change. Many lesson learners may not have fundamentally changed their foreign policy *positions*, but as a result of the war they have begun to reexamine their basic *assumptions*. An editor of an intellectual journal is typical of this group:

Yes, I'm more dubious about a wide set of policies that we continued with probably into the late fifties and early sixties that really we didn't have to. In other words, for example, I think that to hear Kennedy's inaugural today would be a rather shocking experience. I mean, it would sound almost bellicose and, I have a feeling, a bit old-fashioned.[6]

Unlike intellectuals advocating the limits of power doctrine who question whether Americans are in fact willing to "pay any price," those who question the doctrine of the international Communist conspiracy or its accompanying aggressive United

States posture now question the very morality of Kennedy's tone.

Changes in Views on the Cold War

The reason only half the intellectuals interviewed said they had really learned anything from the Vietnam War is that half the sample did not change their views of the Cold War from those they had held in the late 1950's. It will be recalled that one quarter of the intellectuals interviewed claimed that even before Vietnam they had strongly opposed the Cold War. Some felt that Communist dangers to the United States interests were vastly and deliberately exaggerated; some felt the United States was attempting to construct an American Empire. Others—pacifists, anarchists and some democratic socialists—for various reasons opposed both United States and Soviet Union policies. For those Cold War opponents, events in Vietnam hardly offered much reason for change. A radical intellectual explained,

> I think Vietnam was a particularly malicious instance of a general policy . . . of trying to maintain control of the underdeveloped societies of the world.

When asked whether he thought this policy could be changed, he explained:

> It could only be changed by a very radical transformation of American society which would make it possible to use resources for constructive purposes, instead of for maintenance of American power. But that would mean a large-scale sort of revolution.

Far from being pleased that their point of view had been vindicated, many of the old anti–Cold War intellectuals answered our question as to whether the Vietnam War had changed their views of the Cold War in the despairing manner of this culture critic:

I want to answer that carefully. No, I'm afraid at the moment all it's done for me, as it's done for lots of others, is to make us feel very bitter, because it's proving that the things that we as political amateurs predicted would happen have happened, both in the so-called Cold War, itself, and this particular Hot War. And keep happening, and repeat and repeat.

Only one of the early opponents of the Cold War had any other opinion in 1970. Only a few of these anti–Cold War intellectuals felt that they had learned anything as a result of the Vietnam War; and most of these simply felt that the Vietnam War had reinforced their previous views. Perhaps the strongest argument against the Vietnam War's having had any radical effect on American intellectuals comes from the fact that all four (and only four!) intellectuals in our sample who held radical views on the Cold War in 1970 held those same views *before* the war in Vietnam.

Former Cold Warriors, roughly 60 percent of our sample, were also dissatisfied with the war in Vietnam, but as we pointed out, mainly because it did not work. Despite their criticism of the war, about one third of the old Cold Warriors have not appreciably changed their basic views on foreign relations. A conservative novelist who wished that we had used more force in Vietnam replied to our question about change of mind on the Cold War:

No, not basically. I think that every indication is that Communism is an outward moving force and it will continue to be so. I think it is a pattern it has always followed and will continue to follow as long as they can get away with it.

Thus, between unreformed Cold Warriors and the old opponents of the Cold War policies of the United States, over half our sample of leading intellectuals had *not* changed in any appreciable way on the Cold War issue. To be sure, almost half the sample has changed, and the change is all in one direction — toward a more liberal, more flexible yet at the same time less coherent foreign policy. But, as we shall see, change does not necessarily mean change *because* of the war in Vietnam.

Those intellectuals who did change their minds on the Cold War fall into three rather distinct types: the first was a total convert to a firm, anti–Cold War position; the second was a person whose original leftist tendencies stifled by the Joseph McCarthy era were moderately reawakened by opposition to the war in Vietnam; and the third type was not especially ideological one way or the other but has been convinced by the lessons of Vietnam that something new must happen.

There were only 15 in our entire sample who were totally converted by the war into firm anti–Cold Warriors. The following respondent tells the story more eloquently than we possibly could.

> At the beginning of this period [Vietnam war], I was an ex–Naval Reservist with service in two wars behind me – World War II and the Korean War, occupying a position somewhat further to the right on the role of the military as an arm of U.S. foreign policy than the position held by most of my more liberal friends in the academic community. This position was naturally reflected in my attitude towards Vietnam and our engagement there. As the years have moved on and our problems in Vietnam have remained with us, and as the future has continued to yield little clear promise of reconsideration, I have to admit that America must come to a degree of accommodation with the elements of Communist power. . . . There was a time when I thought that we had to look to a final confrontation between the Communist and the non-Communist world. . . . I now find myself quite unpersuaded of this. . . . I certainly remember that I was an adherent of the domino theory. I was also an adherent of the proposition that we must contain Communism in the world at almost all costs. These are positions that I have very radically altered with these Years of Vietnam. . . . *I've changed my mind about the dangers of Communism.* [Emphasis supplied] I've come to feel that Communism almost defies present definition because it varies so widely from country to country. . . . It very well might be that a form of government that we might label Communism, as we now use the term, might very well prove to be an effective economic instrument for bringing a country rapidly into the modern economic era.

His change of "mind about the dangers of Communism" should be underscored, because this is a crucial element in any drastic revision of ideas about American foreign policy.

"Reactivated moderate leftists" were also relatively few in number (17), but they had the highest prestige among intellectuals of any of the types we are discussing, and so, they carried more weight and made a greater impression than their small number might suggest. They differed from the first group in that their break with the Cold War, a term they do not like, came in the late fifties. Unlike the group that always opposed the Cold War, the moderate leftists at one time strongly supported it and this, plus their early support of Diem as a democratic force in Asia and their rejection of revisionist Cold War history made them a special target of radical leftists. Most were democratic socialists. The following quotation summarizes their position.

> On the whole I've been a supporter of the American position taken in the Cold War generally, but I have to go back a very long time to find when my support was anything but severely qualified. . . . The world has changed to an extent that I can no longer find much relevance in what's come to be called the Cold War ideology. The very term's a kind of put-down, since ideology is now used to describe the ideas of people with whom you don't agree. . . . I think the American response from roughly '45 to the early '50's was in many ways a creative and justified and inevitable one. . . . The ascendence of Soviet power . . . may very well have been exaggerated, but it was very hard to gauge its proportions at the time. . . . Maybe we overreacted and maybe we didn't. On the other hand, to say we did is to take advantage of hindsight – it was not so obvious at the time. But I'm only talking up to about the 1950's. Other things I did oppose at the time, such as SEATO, and the Baghdad Pact, the landing of the Marines in Lebanon, and most of all, the American intervention, bar none, in the Western Hemisphere: Guatemala, Cuba, Dominican Republic – I never supported any of these for one moment. In fact, I was and I am more shocked by the Dominican intervention than by what happened in Vietnam.

Reactivated moderate leftists had, indeed, changed their views on the Cold War, but they themselves did not attribute this change to the war in Vietnam. Most, by the way, were more militant on the Vietnam issue than the respondent quoted.

While cognizant of the dangers of Stalinism, none thought Communism was monolithic, at least not after the death of Stalin. We call them "reactivated" because almost all of them were either former leftists or current socialists who were once strongly opposed to any kind of American intervention (some even opposing World War II). They were "moderate" because they rejected a radical line, and some were quite critical of current thinking in the intellectual community about American policy.

Finally, there were those who merely learned lessons; neither converted nor reactivated, they simply began to feel that something was wrong with American foreign policy though they were not ready to take a strong anti–Cold War position. The most characteristic lesson they learned from Vietnam was tactics rather than politics: most felt they learned about the limits of American power. In contrast, the Vietnam converts emphasized that they no longer believed in the Communist conspiracy doctrine while the reactivated moderate leftists denied that they had ever believed in it. Almost all the Vietnam converts, as well as almost all of the lesson-learners, switched their Vietnam position from one favorable to intervention to one opposed. Three quarters of the activated moderate leftists had always been opposed to the war; all of the constant anti–Cold War intellectuals were opposed to the war from the start. But a good majority of unreconstructed Cold Warriors still would have liked to prevent a Communist takeover in Vietnam.

Intellectuals with differing patterns of change on the Cold War also differed, as one might expect, in what they did not like about the war. Almost all of those who had always been against the Cold War opposed the war in Vietnam for various ideological reasons; about a third also invoked moral reasons. Almost no one opposed it for pragmatic reasons. Demonstrating their common heritage, reactivated moderate leftists offered reasons for opposing the war similar to those of constant radicals but gave more pragmatic and fewer ideological reasons. Lesson-learners and converts both emphasized pragmatic reasons, but offered a smattering of ideological and moral reasons for opposing the war. Finally, three fourths of even the unreconstructed

Cold Warriors found something to object to about the war in Vietnam, but these objections were almost always couched in pragmatic terms.

The Program for the Future

It is one thing to *say* that you have learned a lesson. Change from an objective point of view may be something else again. In objective statistical terms, the question is how much do we need to know about an intellectual's position on Vietnam to know what he thinks about foreign policy issues today. Almost all the intellectuals who in the late fifties were opposed to the Cold War or even moderate on that issue were in 1970 opposed to future Cold War policies and tactics. So what they thought about Vietnam is irrelevant to an understanding of their 1970 policies. Not so for the strong old Cold Warriors (Positions 1 and 2 in the late fifties). What they thought about the war in Vietnam was very important. For, of the old Cold Warriors who wanted to get out of Vietnam immediately, and there were some, more than half opposed Cold War tactics in the future; one third of those who wished to negotiate also opposed future Cold War tactics; but almost all of those who would still like to prevent a Communist takeover endorsed Cold War tactics in the future, as well. Again, this statistical analysis suggests that much of the change in Cold War orientations had begun *before* the Vietnam War and that relatively few were converted by the war.

To be for or against the Cold War is too simple a position. It also tends to convey a sense of directness and purpose which is lacking in most of our interviews. Let us look at the material itself more closely. To begin with, about half of those for whom we have some sort of opinion about the Cold War in 1970 seemed to be opposed to it. The most radical position is one we have already encountered (on Page 204). These opponents of the Cold War favor aid to revolutionary governments. More basically, they feel that only "changes in the fundamental structure of American society" will produce an adequate foreign policy. Several felt that "the United States would have to recognize that China is going to be the dominant power in the Far East," and that "American policy is a good deal more ad hoc and stupid" than they had thought. There are only four proponents of this position and all are among the youngest in our sample.

A second anti–Cold War position can be called the "new realism" and stems from the desire to confine United States activities to those situations which directly involve only its own interests. This orientation is held by about one third of the Cold War opponents, and is distributed more or less equally among Vietnam converts, reactivated moderate leftists, and anti–Cold War intellectuals.

The last anti–Cold War position, support of democratic governments which meet people's needs, is the most popular. It is also probably the best articulated policy, since it is and has been the party line of the socialists, at least since the late 1950's. In fact, it is this line which led some of our sample to support Diem in the 1950's, a support which seriously alienated the New Left. Although fairly well articulated in its ideology, the position is not detailed in its application. Intellectuals holding this position all favor the support of certain kinds of national Communist or socialist governments and oppose United States support of purportedly anti-Communist, nondemocratic governments. One reason some of the old hands with this view are now so unclear is that Vietnam has made them less certain that their views can be made to work. A socialist who has always been opposed to the Cold War put it this way when asked whether Vietnam had changed his views:

> I'd say yes. . . . To be antiwar but also anti-Communist after all these years. . . . Anti-Communist in the progressive sense, not in the Cold War sense. A lot of young people don't realize that you could be both. . . . And I'd say another thing the Vietnam War may have done. . . . It has emphasized in my own thinking the dangers of being the social workers of the world. So I haven't changed my fundamental perspective, but I'd say it's made me more prudent and cautious.

The line between prudence and caution in pursuing an anti-Cold War policy, and flexibility and caution in pursuing a modified Cold War policy is at times hard to draw. All the lesson-learners had a mixed view toward foreign policy, one which was not clearly against all Cold War policies. They felt that the United States should be more flexible and less bellicose; on the other hand, it should also help friendly governments whenever possible. The use of military force was not automatically ruled out, but it was important to wind down armaments. World Communism was still a danger, and yet we must learn to coexist in a more viable way. Many of the lesson-learners were themselves former leftists, not totally recovered from their bouts with Stalinism, as the historian and critic quoted below indicates.

> I have very mixed feelings, first of all about the Soviet Union, and I suppose my experience in —— in 1951–1952, where I actually was propositioned by a Soviet Secretary who, I afterwards found out, was in the NKVD, who offered me all kinds of things – women, money, all the rest of it – if I would state that the Americans were using germ warfare in Korea. It was this close experience with Soviet practices at that time – and then the year I spent in ——, 1962–63, which again saw the role of the Soviet Union. So that at the same time I recognized that it was an absolute necessity that we worked out some kind of modus vivendi, with the Soviet Union, that's why I think the SALT talks are so tremendously important. And while I have no brief for their society, I think it's a tyrannical state and really simply continuing in the old policies of the imperial Czarist government under a new regime with . . . the use of any kind of technique, including anti-Semitism, to bolster its interests. The fact, I suppose, is that it's not so much ideological; it's the feeling that I have all over the world that there's a certain breed of man coming to the fore and that the Soviet variety of the bureaucrat is simply more dangerous and deplorable because the Soviet style of bureaucrat has more power.

The prescriptions for action suggested by this respondent and others of this type are notably vague. Rather, a mood is conveyed. Whatever it is, there is no clear abandonment of a Cold War orientation.

Advocates of a basically unreformed Cold War point of view, whom we have already met, were much more coherent in their policy statements, though some suggested that intervention tactics needed to be changed.

> Vietnam has certainly proved to me that we need to know a lot more about ways of countering guerillas than we do know or have known. . . . I think there are places where we might have to go in, there are places where it would be better if we went in indirectly. . . . We might have to go in, but I prefer to do it with volunteers, rather than with conscripted soldiers.

The lessons of Vietnam were most easily applied to Southeast Asia and even unreconstructed Cold Warriors tended to be less willing to engage in further military defense in this area. But even in this specific situation, when pressed for longer-range planning than the immediate Vietnam situation, one quarter of the respondents did not know, could not respond or were evasive. The most popular long-range policy toward Southeast Asia, outlined by almost half of those taking a position, was to have no policy or to leave countries in this area alone to do their own thing. For example, the following anti–Cold War reactivated moderate leftist was annoyed by our question on Southeast Asia policy.

> I don't see why we should have one [a policy in Southeast Asia]. We don't have a policy in Central Africa that requires this amount of soul-searching or tension. We obviously have a policy there, even if we have no policy or a policy of inadvertence. . . . If you, not you but we, if we insist on asking that question, then we are going to be forced into making much more elaborate policy than we need. . . .

Others wanted a specific policy of neutrality.

> We should eliminate all our armed forces in Southeast Asia, and eliminate all American bases on the continent of Asia . . . and in general, make the main thrust of our policy in the direction of a neutralized Southeast Asia.

Laissez-faire for Asia might include, said a professor of political science, "admitting that perhaps we have little to offer there in the way of a model of government or the economy. . . . We may have to be prepared to accept the fact that there may be governments in power which are not capitalistic . . . [but] which may be the best thing from the point of view of those involved."

The second most popular orientation, favored by over one third of the intellectuals, proposed some kind of American aid or involvement of a nonmilitary nature. A typical proposal was:

> I think it [our policy] should be one of continuing social-economic aid to whatever nations need it. . . . I don't think we can turn our backs on Asia and Southeast Asia. . . . We should cooperate more with the Japanese economically, and extend aid to all of Asia including, hopefully, North Vietnam, which obviously needs a lot of rebuilding. . . . I don't believe we should make any further military commitments in Southeast Asia, no.

Some respondents with this economic aid orientation also qualified their nonmilitary approach by saying that if India or Australia were attacked or strongly threatened by China, then we ought to give military aid.

Even intellectuals with more of a Cold War orientation hedged the nature of possible military commitments. One approach was to be vague in applying existing Cold War policies.

> Same as the policy in Europe and the Mideast. That is to say, a policy of trying to establish and maintain tacit understanding of the Soviets and the Chinese involving the renunciation on their part of attempts to change political and social situations in separate countries by force [that is] external support for internal subversion, or conquest in any other form.

Another emphasized American military presence without specifying the circumstances under which the armed forces might be used. Only 15 percent of those offering policy suggestions on Southeast Asia held similar opinions. Almost all of these were unreconstructed Cold Warriors in their general foreign policy orientation. As might also be expected, almost all the converts were opposed to any American Southeast Asia involvement, but a third of those who always opposed the Cold War were willing to consider economic but not military aid, a position favored by more than half of the activated moderate left and the lesson-learners.

In short, Southeast Asia remained something of an enigma. Even the relatively few remaining ardent Cold Warriors did not uniformly endorse a policy of military intervention. On the other hand, except for recent converts to an anti–Cold War policy who tended toward an isolationist posture in Southeast Asia, many of the other respondents were willing to consider some sort of American presence short of military aid. When it comes down to actual cases then, the doctrine of Asian exceptionalism seems to have been held by more intellectuals than cared openly to admit it. The result is that while Vietnam has made some few genuine converts to a general anti–Cold War policy, it has produced a much stronger neo-isolationist position with respect to Asia itself. So, in answer to the question, has Vietnam produced among intellectuals a major reorientation in foreign policy, we can conclude with assurance only that hopes of building American-style democracies in Asia have all but entirely faded and that even in 1970 coexistence with China would have received strong support. As for Europe, Latin America, the Mideast, and Africa, a considerably larger proportion of intellectuals than in 1960 would probably oppose American military involvement, but many things other than Vietnam have contributed to that stance.

Conclusion

The American intellectual elite were overwhelmingly opposed to the foreign policies of their government in 1970; despite some claims to the contrary, there is little reason to believe that the situation in the second Nixon administration even with the Kissinger gyrations will be very different. We have been at great pains to point out that the actual beginning of this rift between intellectuals and their government in the present era did not occur in the middle sixties when the war in Vietnam became a major issue, but in the late fifties when the old Cold War formulae seemed to many intellectuals no longer appropriate.

A sense of unease needs something dramatic to help it jell into outright opposition, and a series of blows — Cuba, the Dominican Republic invasion and, most of all, the Vietnam War served this function. Events alone, however, do not make for opinion; this is as true of the intellectual elite as it is of the general public. Events, especially those on the other side of the world, first need to be called to one's attention; then they must be interpreted. A major reason for the intellectual community's early opposition to the war in Vietnam was not only that the late fifties had begun to prepare the ideological and intellectual groundwork, but that the structure of the intellectual community included within it the antennae which allowed it to become quickly sensitive to the war.

Some say, as we noted in Chapter 3, that there is no such thing as an intellectual community in the United States. There are only communities and there is also much isolation. While this may be so for some purposes, and while there are certainly identifiable subcircles in the intellectual community, when a "hot" issue like Vietnam comes up, we have shown that almost everyone in the community has a line to at least one major original source of information and interpretation. In this respect, there was an intellectual community on the Vietnam issue and it functioned very much like a small town in spreading the word. And the word crisscrossed from one circle to another so that most political and ideological boundaries became obliterated.

One of the factors that has been said to lead to fragmentation and isolation among American intellectuals in the case of the Vietnam War led, on the contrary, to integration. American college and university campus life has been blamed by many for the lack of an apparent center to American intellectual life. But on the Vietnam issue, campus militancy at the leading universities, where most of the leading intellectuals who are on campuses are located, lent an appearance of unanimity. At the very least the campus teach-in movement brought into the Vietnam discussion those intellectuals who had not previously been much affected or those isolated from the center of the discussion. In the end, as we shall see in the next section, the American intellectual elite resoundingly rejected the counterculture movement, which had joined forces with the anti–Vietnam War movement. Yet in their heyday from 1965 to 1967, these campus movements brought along some of the last of the holdouts.

We had talked earlier about the intellectual elite's generalized expertise. But each intellectual also has a specialty, and the community of general experts systematically draws upon the specialists within the community or brings in outside specialists when a particular issue makes that necessary. These issue experts then may prove so attractive to other intellectuals that they remain within the community, becoming themselves generalists on other topics. Not only then did the network function to spread the word, but the expertise was already available to be drawn upon without too much extra effort and this, too, helped spread the ideas before other groups in America had caught on. Often this expertise first appeared in print, but our network data show that the intellectual mass media has much more of the air of a small-town newspaper where the reporter is known to everyone than the aura of contemporary national mass society media in which the stars are images on a tube. Once the experts started the ball rolling, of course, some of the leading intellectual generalists joined the debate.

The structure just described explains why elite American intellectuals were not radicalized by the war in Vietnam. To be sure, they became more and more vehement on the war as the years went by, but as we have pointed out, their criticism was more pragmatic than ideological and, except in later years, hardly moral. Certainly very few were radical. The reasons for this are straightforward. The word was first spread from a core of nonradicals, for the most part. The channels followed the existing informal institutional patterns of the intellectual community as it existed in the late fifties. These channels and structures – the social circles, the journals and the rating system described in the first several chapters of this book – were oriented to Cold War liberalism and especially to an anti-Communist liberal point of view. In order for basic orientations and basic values to change, it is necessary that the basic social structures change, as well. Now we are not arguing here which comes first – whether values or structure are epiphenomenal – but merely that for one to change, the other must also change. But it is very clear from our analysis of the structure of the intellectuals' circles and the structure which surrounded discussion of the war in Vietnam that during the sixties there was no radical change in that structure. The strains of the Vietnam War did loosen circles and relations but the networks were not totally restructured. The leaders were still moderates and liberals. Drastic change in the formal patterns – the journals – and corresponding change in the informal networks would have had to accompany drastic changes in values.

There were important changes in the sixties but they did not permeate the elite intellectuals in a fundamental way. The most important change was a resurgence of a more powerful and pervasive counterculture. But the counterculture, as we shall see, was rejected by almost every one of the leading American intellectuals. Counterculture heroes, "new journalists," new filmmakers and critics by and large had little contact with the set of intellectuals we are talking about. The upstarts may have said they were not even trying to break into this group – but whether they really tried or not, the fact is that they were not permitted, for the most part, to enter. The counterculture did affect some of the New York literary circle, but not profoundly, and certainly had little effect on the social science–literary circle. Youth would not be served.

The counterculture was radical in its views on Vietnam. But there were also politically radical circles quite divorced from the counterculture. A circle of such radicals did enter the intellectual elite, though because it contained older intellectuals who had been reradicalized by the sixties, the average age of the radical circle was not much different from the age of the rest of the intellectuals. Though the circle of radicals was much published, our data show they remained isolated.

There was one important structural change in the sixties: the appearance of the *New York Review of Books,* an appearance we discussed in Chapter 2. The *New York Review of Books* published some of the most widely read strong, moral and ideological criticism of the war, some of it from a radical point of view. Though the *New York Review* was the journal most widely read by the intellectual elite in the late sixties, for a variety of reasons it failed to create around it a true, identifiable network. As we pointed out in Chapter 3, there is no special *New York Review of Books* "crowd." Without such a circle, the chances of a strong ideological impact of the *Review* on leading intellectuals was small, indeed. On the war in Vietnam, the *New York Review of Books* may have had much more impact on the more isolated college and university youth markets than on the more central elite intellectual circles.

We have seen that the American intellectual elite failed to become radicalized by the war, that their objections were mainly pragmatic and that their lack of basic change, while at the same time their change in opinion toward an object — Vietnam — took place, was in part because their social structure did not change, too. Why does this matter? Because, for a brief time, intellectuals were listened to by an important segment of America — liberal politicians and college youth. In 1973, twelve years after the war began, Nixon was apparently finally able to put an end to the war and claim he had reached his objectives — peace with honor. McGovern, whose avowed aim for years was to put an immediate end to the war, was unable to convince America that he stood for peace in Vietnam. Why? Because the overwhelming majority of Americans opposed the war, not because it was wrong from a moral or an ideological point of view, but because we did not or could not win.[7] In this way, the public was not basically different from the intellectual elite. Had the intellectuals abandoned their pragmatic *realpolitik* and attempted to "sell" the public, either directly or through a mass media "trickle down effect," contemporary history might have looked very different, though, as we shall see from Part IV, this view may grant far too much influence to the intellectual elite. While it is possible that intellectuals took the stance they did merely in an effort to appeal to policymakers and to the public, the fact is that they were merely following their own inclinations. Both the strategy, if any, and the inclination itself may well have backfired.

III

American Social Problems and the Intellectual Elite

9

What Is Wrong with America

We come now to matters that lie at the core of the intellectual's role as moral and social critic. The war in Vietnam was of course a special case, a crisis situation which mobilized the interest of almost all intellectuals. In the normal course of events, if indeed any decade of the twentieth century can be considered normal, intellectuals tend to be interested in a variety of different problems. They tend not to specialize, shifting their interests from time to time and talking about problems that are not necessarily within the direct sphere of their technical competence. Yet at any one time there is a division of labor with respect to social criticism within the intellectual community which corresponds roughly to differences in the way problems are perceived, as well as to differences in personal concerns based on an intellectual's own history. Though intellectuals tend to work on social problems which differ from those which occupy the rest of the American elite, the great surprise is the extraordinary similarity among the intellectuals, the elite as a whole, and the general public in what they see as wrong with America.

This section is divided into six chapters. The first describes what intellectuals, other elites, and the public each see as wrong with this country, and discusses briefly how intellectuals talk about these matters. The second chapter shows the areas of social and cultural criticism of current concern to intellectuals. Chapter 11 takes up the "classic" issues of foreign policy and domestic reform; Chapter 12 deals with the confusion among the remnants of the once positive and persuasive exponents of racial integration; Chapter 13 appropriately follows by reviewing the emergent issue of "culture crisis"; and Chapter 14 briefly concludes that, like everyone else, intellectuals have no definitive answers to the crises of our civilization, but specialize, rather, in asking questions.

In some ways the interview technique used here is unfair to intellectuals, most of whom tend to express themselves best in writing. Though rough edges are revealed, the interviews do show how people think on their feet and in the case of intellectuals this is very important to know. I must confess to a feeling that the portrait which emerges here is a harsh one and unflattering at best. But the sixties have left intellectuals in a terrible quandary and it would be unfair to them if their predicament were in any way concealed or whitewashed. For it is out of confusion that intellectuals best create a new synthesis.

Intellectuals were asked at the beginning of all but our short and telephone interviews: "What briefly do you think have been the most important problems facing the United States in the last twenty-five years?" A bit aghast at the magnitude of the question, most respondents paused for breath, noting that the period covered was roughly since the end of World War II. Then, after brief reflection most launched into an extended discussion, which in many cases had to be cut short so that we could proceed with the rest of the interview. And it was in this context that they picked the one problem with which they had been most concerned in the past five years. They were also asked, "What problems do you think will be important five years from now?" The answers are classified in Table 21.[1]

In view of the general situation of the United States after World War II coupled with the then (1970) current Vietnam War, it is no surprise that foreign policy leads the list of problems. Problems in race relations are second. After these two,

TABLE 21

American Problems as Seen by the Intellectual Elite, 1970 (By Percentage)

	PROBLEMS OF THE LAST 25 YEARS (N = 91)	PROBLEMS 5 YEARS FROM NOW[a] (N = 70)
Foreign policy	82	67
Race	66	61
Culture and values (Culture crisis)	34	45
Economics	31	25
Social issues	24	32
Ecology	20	36
Government organization	12	6
Law and order	8	11
Urban problems	2	21
Other	18	3

[a] Including the problems noted by those who said "Same as today."

seen as major blights, culture and values and economic problems are next, followed by social reform and ecology. Government reform, law and order are next, and urban problems, really covered by many other categories, comes last. Showing the inability of intellectuals to fit into neat categories, the category "other" contains an astounding 18 percent.

How unique is this view of America? Do intellectuals, because of their involvement with the war in Vietnam and with culture and values, tend to overrate these problems as compared with other members of the American elite or the public at large? Hardly. The American intellectual elite have almost the same views of the problems facing the United States as the general public, and differ from the rest of the elite mainly in their greater concern about race relations.

TABLE 22

The Most Important Problems Facing the Country by Institutional Sector (Ranked 1–9)

	PROBLEM									
SECTOR[a]	FOREIGN POLICY	RACE	CULTURE & VALUES	ECON-OMY	SOCIAL REFORM	ECOL-OGY	GOVT. REFORM	LAW & ORDER	URBAN PROBS.	NUM-BER
Economic										
Business	2	4	3	1	5	7	6	9	8	(130)
Labor	2	5.5	4	1	3	5.5	9	8	7	(48)
Political										
Administration	2	4	3	1	6	5	7.5	9	8	(113)
Congress	1	8	6.5	2	3.5	3.5	9	5	6.5	(55)
Party	2	3	4	1	5	7.5	9	7.5	6	(50)
Voluntary Association	1	4	3	2	5	8	9	6	7	(50)
Symbolic										
Mass media	1	4	3	2	8	5	6	9	7	(62)
Intellectual	1	2	3	4	5	6	7	8	9	(110)
The Public[b]	1	2	3	5	7	6	–	4	–	

[a] Intellectual sector based on Table 21, 1970. Other elite sectors from the American Leadership Study, Allen Barton, Bodgan Denitch, Charles Kadushin and Carol Weiss, study directors. Leaders were asked for the "three most important problems facing the country today" (June through December, 1971). The question is obviously time-sensitive, reflecting the concern with economic problems of that time.

[b] The public ranks are based on average responses to Gallup Poll questions, asking for the most important problems facing America. Responses were averaged by Funkhouser, "The Issues of the Sixties: An Exploratory Study in the Dynamics of Public Opinion," *Public Opinion Quarterly,* 37 (Spring 1973):66, for the years 1963–1970, thus providing a fairly good comparison with the intellectuals who were asked for an extended time period. The Gallup categories are slightly different from those used in the elite study. Foreign policy is represented by "Vietnam," Culture and Values by "Campus unrest," "Polarization," and "Declining Morality." Law and order is represented by "Crime and Lawlessness" and "Drug Addiction." Economics refers to "Inflation," and Ecology to "Pollution." Finally Social Issues are represented by "Poverty."

The bases for these assertions are varied. In the summer, fall, and winter of 1971, the Bureau of Applied Social Research Study of American Leaders, of which this study of intellectuals is a segment, asked over 500 top American leaders in different fields, "What briefly are the three most important problems facing the country today?" The emphasis on today rather than the past twenty-five years and the fact that three problems were requested rather than the five or so mentioned by intellectuals makes their data a bit different, although the same categories are used in both studies. The data for the public are based on an analysis of Gallup Poll questions by G. Ray Funkhouser[2] in which he averaged the poll results for the years 1963 through 1970 – thus making them somewhat comparable to the data in the intellectuals' study which covered the recollection of twenty-five years. Table 22 shows the rank order of the various issues for the different elite groups and for the public.

The inflation–wage price freeze affected the perceptions of American leaders interviewed in 1971. Based on the number of leaders who said a given problem was one of the top three facing the country, economic problems were either first or second in importance for all sectors of the American leadership interviewed in the second half of 1971. Although the economy rated only in fifth place for the public during the sixties, in July 1971 Gallup Poll results showed it was perceived as the top issue for the coming presidential campaign. The economy rated fifth among intellectuals who were interviewed in 1970. Very few were at that time concerned about inflation; rather, their concerns were about the basic nature of American economic institutions.

If Nixon's dramatic conversion to controls over the economy is taken into account then foreign policy is the number one problem for all elite sectors, and the number one problem for the public as well. Race, the number two problem for both intellectuals and the public who were reflecting the events of the sixties, was downgraded in 1971 by most sectors of the American leadership to fourth place. This may reflect the fact that there was little in the news about race and race riots in 1971. The public too in July, 1971, rated race close to the bottom of the list as a likely issue in the 1972 presidential campaign. Yet the prominent place given to racial problems by intellectuals is not wholly an artifact of a narrow contemporary perspective. For in 1970 intellectuals tended to believe that problems in race relations would be even more important in five years. And there were no major race riots in 1970, though the New York Times Index did show about 25 percent more listings on "Negroes" in 1970 than 1971.

Perhaps the most surprising aspect of Table 21 is the general agreement among the public, the American leadership as a whole, and the American intellectual elite that the area of culture and value problems is of major importance, and ranks third only behind foreign policy and economics or race relations. Intellectuals and the rest of the leadership talk somewhat differently, of course, about value and culture problems. The American leadership as a whole tends to think of the problem more in terms of "moral decay," while the intellectuals are more concerned with alienation, the "culture crisis," and the crisis of the university. But the problem area is definitely the same one. The categories into which public concern about values fell, however, were much the same as those of the intellectuals.

Despite this picture of general agreement among the various sectors of American leadership on what is wrong with America, there are some important differences. Intellectuals, as we said, give more prominence to race relations. Labor and business differ, as one might expect, on the priority given to social reform, with labor leaders more likely to see problems in this area than leaders in any other sector, including intellectuals.

Congress, on the other hand, differs from all other sectors in important ways. Their priorities seem to be given to those issues which are actually before them or which they feel Congress might do something about. The impotence of Congress with respect to foreign policy must bother them, for almost 80 percent of the congressmen in our study said foreign policy was one of the three major problems facing the country — 20 percent more than the nearest other sector (excluding intellectuals). Congress was also more likely to rate social reform and ecological issues as major problems. These were areas in which Congress was then considering major legislation. Race relations, by contrast, were downgraded because Congress has almost always been blocked from action in this area by the veto power of Southern senators.

There is one major difference between *all* the elite, except Congress, and the general public. The public is much more concerned about law and order than any sector of the elite. Perhaps the elite is sheltered from some of the "crime in the streets problem" which concerns the public and perhaps the elite is also less concerned with drugs as a personification of evil.

But there is one other important explanation for the greater concern of the public with law and order: what they read. And this may also explain the great similarities between the various sectors of the elite. In a very ingenious demonstration, Funkhouser suggests that the issues which the public sees as problems facing America parallel most closely their coverage in the news rather than their existence in "reality." This coverage is according to "newsworthiness" rather than in proportion to the objective importance of the problem. And it is what is in the news that influences the public. The nation's elite all read the same newsweeklies and newspapers (the *New York Times* and the *Washington Post* lead). It takes only one day's glance to conclude that the *Daily News* has proportionately more coverage of crime and more dramatic coverage of it than the *New York Times.* So it may be not only where the elite live in comparison to the general public but also what they read that influences their perceptions of law and order as an issue.

To return to the intellectuals, the frequency with which they mentioned problems does not necessarily correspond to the intensity with which the problems were felt. One way of assessing the importance of a problem to the respondent is to see whether it is the very first thing he thinks of when asked the question. Thus 60 percent of those who mentioned foreign policy at some point in the discussion of problems facing the United States mentioned it as the very first problem. The closest rival in saliency was race, with 33 percent of those who talked about it mentioning it first. Economics was next with 29 percent saliency and government reform, surprisingly, was given top billing by 27 percent of those mentioning it. In contrast, though the culture crisis was frequently mentioned, it was rarely mentioned first, with only 12 percent giving it such a prominent place. Other issues were even less likely to be mentioned first. Foreign affairs thus remains both the most frequent and the most salient issue, while something as attractive to intellectual interest as culture is recognized as being of somewhat less overall fundamental importance than the Cold War or the war in Vietnam.

The problems facing the United States have been with us for at least twenty-five years, say most intellectuals, and about 70 percent of those mentioning a specific problem area feel that it will still be a problem five years hence. Though most thought that the Vietnam War would soon be over (it actually took three years for it to be officially declared over), most also saw foreign policy as remaining problematic. Race relations too were felt to be a continuing issue. The culture crisis was seen as becoming even more important, as were issues of social reform and ecology. The crisis of cities is especially singled out: only two percent specifically noted it as a problem of the last twenty-five years, but more than 20 percent picked it as likely to be important five years from the time of the interviews. Now that three years have passed since the interviews, these prognostications still seem reasonable.

The characteristics of an intellectual affect to some degree what he sees as American problems, though the differences between intellectuals of one camp and another are generally not overwhelming. For example, foreign policy is seen as a major problem more by older intellectuals who have been around to see the twists and turns of American policy since the 1930's.[3] And because the social science professors in our sample include a fair number of foreign policy experts, they tend naturally to see foreign policy as a major problem more than do members of other professions.

Race, in contrast, is a problem more often seen by younger intellectuals, who, as we shall see, became especially concerned about race in the fifties and early sixties. Jewishness (but only among nonacademics) and prestige among intellectuals (but only among non-Jews) is associated with perceiving race as a problem. Eighty-five percent of the humanities professors saw race as a problem, and this reflects a combination of historians who have written on race, and the English professors who have been impressed by Black writers. But these findings (except for the occupational one) are mainly a reflection of the correlation of "liberalness" and general political activity (especially on Vietnam) with seeing race as a problem. In this connection we must sadly observe that intellectuals who had the very best regular channels to Washington in this period were less likely than intellectuals without this access to see race relations as a problem. Benign neglect, indeed!

Culture and values are the problems most often seen by the most prestigeful intellectuals who, as we shall see were especially aggrieved by the counterculture and by alienation in America. Signs of prestige in our study include the number of books reviewed, the number of articles written, the number of mentions by other intellectuals both on Vietnam and not on Vietnam and, of course, membership in the center circle. In addition to the obvious factor of prestige, age, Jewish background, and an academic occupation are all related to seeing problems in the area of culture and values, and these factors hold true, we remind the reader, even when controlled for each other and for prestige. All these backgrounds or characteristics have an investment, of one kind or another, in high culture and in the American values of freedom, hard work and success. Now intellectuals who see culture and values as a problem felt all these values were being eroded by the counterculture and youth.

Seeing economic issues as a problem is partly a function of where one lives — New Yorkers seem more exposed to the bourse and to issues of poverty — and this is especially true of nonacademics in New York. Radicalism, as measured either by circle membership or by self-designation, is also associated with an interest in economics. But the economy is also a concern for technical economic advisors. Those intellectuals with good regular channels to government were especially likely to see economic problems as important.

Social reform, too, was a problem more perceived by those who were more liberal or radical, and/or who were Vietnam activists.

Finally, ecology is the problem seen by the "squares" in intellectual life – the conservatives, nonacademics living outside New York, and non-Jews.

In summary: the American intellectual elite see roughly the same things wrong with the United States as do other Americans and the rest of the elite except that their greater concerns with civil liberties and civil rights may have made them more sensitive to race problems. Especially noteworthy is that, despite their greater personal concern with matters of spirit, intellectuals are no more likely than others to rate culture and values as important problems. All sectors and the public rate this issue as important. But law and order, an issue of great concern to the average American, does not rate either among intellectuals or the rest of the elite; if it is mentioned, law and order is usually integrated, as we shall see, into the issue of "culture crisis."

We were concerned not only with what intellectuals thought was wrong with the country but with how they generally expressed themselves. One thing is especially clear about our sample of intellectuals: they eschew technicality and take a broad humanistic point of view whatever their own occupational specialty. Less than one fifth used any kind of technical "jargon" in talking about problems facing the nation and issues of concern to them. Less than 15 percent invoked any systematic theory, though radicals were somewhat more prone to this style. A seemingly value-neutral, factual manner of presentation was also characteristic of the large majority. Only slightly more than one quarter of the sample made explicit value judgments in describing America's problems – though to be sure an implicit judgment was almost always present. Members of the social science–literary circle were particularly circumspect in voicing value judgments, a reluctance which may follow from the interest of many members of this circle in technical solutions to policy problems. In contrast, members of radical or literary circles were much more likely to offer explicit judgments as to the good or evil of current problems and policies designed to alleviate (or produce) them.

The absence of jargon, theory, or explicit value judgment did not mean, necessarily, that intellectuals eschewed overt and explicit ideological references to liberalism, Communism, social democracy or the like. Forty percent of the sample made such references in talking about problems and issues. Intellectuals who were not circle members tended less often toward explicit ideology, with only 25 percent mentioning a specific "line." Ideology has traditionally been important currency among intellectual circles. Characteristically, intellectuals attack ideologies other than their own with an even slightly greater frequency than they mention ideologies in a favorable context. More than half the sample attacked one ideology or another, with the New Left and radicals bearing the brunt of the attack.

Ideology and style bring us to a more personal view of issues, events, and social problems. For it is one thing to say that from an objective point of view the country has certain problems and quite another to say that you personally are concerned with them and working upon them. The public, the American elite as a whole, and the intellectuals all exercise personal concerns that are at variance with their view of the country's problems. The rest of this chapter is devoted to the issues which intellectuals themselves were concerned with and were working on.

Concerns with issues have a history. Rarely do they spring up full grown, and this is probably more true for intellectuals than for other groups in the population. After our respondents had talked about the most important problems facing the United States in the past twenty-five years, they were told:

> Here is a list of issues split into 5 year time segments. For each time period, indicate what issues you were most concerned with at that time.[4]

Table 23 shows for each time period the problems intellectuals ranked as either first or second in importance to them.[5]

Foreign policy has remained the chief personal concern of intellectuals in the entire period since World War II, matching

TABLE 23

First or Second Issues of Concern to Intellectuals 1945–1965 (In Percentages)

ISSUE[a]	YEAR 1945–1950 (N = 76)	1950–1955 (N = 81)	1955–1960 (N = 80)	1960–1965 (N = 80)
"Foreign policy"	61	66	59	65
"Arms race"	49	39	34	24
Domestic reform				
"Social welfare"	26	13	21	19
"Labor and unions"	9	7	8	1
"Education"	9	10	11	15
"Poverty"	4	5	4	5
"Economy, capitalism"	4	1	1	2
"Race relations"	15	22	42	56
Culture and values				
"Domestic Communism," McCarthyism	10	21	2	1
"Mass culture," culture revolution	8	8	7	1
"Other"	4	1	1	2

[a] Issues in quotes were those actually presented on the list.

their perception of foreign policy as the main issue facing the country. Our strategy in focusing first on the war in Vietnam is confirmed not only as expeditious for the late sixties but as stemming from the almost permanent concern of intellectuals. A specific aspect of foreign policy, the atomic arms race, seems to have steadily declined in interest since its peak right after World War II. And the decline does not seem related to the ban on testing in the atmosphere since there was a ten percentage point decline in interest even in the early fifties. The area of domestic reform seems to have held up as an area of steady concern throughout this period, though the specific issues have changed somewhat. Labor and unions as a matter of personal concern to intellectuals practically disappeared in the period right before our study, while concern with education has increased. McCarthyism shows a predictable peaking in the years 1950–1955 and by 1965 almost no one was personally concerned about either McCarthyism or domestic Communism. And the slogan of mass culture also declined as a topic of interest by 1965. I believe we did not find a right checklist word for what later became a matter of great concern – the so-called "culture crisis." Neither McCarthyism nor "mass culture" properly expressed the current formulation. It may be, however, that this concern truly lessened after 1965.

The issue which shows the most interesting trend is race relations. Starting from a relatively low point of 15 percent checking it as either their first or second social issue of personal concern in the five-year period right after World War II, race relations zooms to 42 percent in 1955 to 1960 and places second only to foreign affairs with a 56 percent "vote" in the five-year period 1960–1965. This increase in concern coincides with the burgeoning of the civil rights movement in the late fifties, a movement led not by the white intellectuals in this sample but by Black leaders. And it was not the established intellectuals who were most affected by the "movement" but the younger ones. For example, by 1960–1965, 84 percent of those in our sample who were under fifty in 1970 checked race as either their primary or second concern in that period; 45 percent of those fifty to fifty-nine did so, and only 33 percent of those sixty years of age or older checked race relations.

The picture of American intellectuals as of 1965 simply does not prepare us for some of the dramatic shifts in attention that occurred in the next five-year period and which produced trends which are still very much with us. Yes, foreign policy was a great personal concern to intellectuals and this matched their objective view of what was wrong with America. Race relations seemed well on its way to becoming the "hottest" issue, even surpassing the war in Vietnam. Domestic reform seemed not to call out a majority of intellectuals but there was always a solid interested minority. However, the continuation of the war led to certain internal crises in the United States which changed this entire order of personal priorities. By 1970, as we shall see in the next chapter, a wholly new set of concerns had developed.

10

The Concerns of Intellectuals

Seventy percent of American intellectuals at the time of our interview reported that the war in Vietnam was the issue of most concern to them. But this does not mean that intellectuals were exclusively concerned with the war. After the main problems facing the country had been discussed, we asked each person interviewed to select one issue other than Vietnam with which he himself had been most concerned over the past five years. After some fencing, most respondents managed to focus on one area or complex; a few settled on two topics. The problem was not that intellectuals had few interests but that they had so many. As a result of this probing we can say that four areas other than Vietnam occupied the attention of American intellectuals during the late sixties: the culture crisis, race relations, domestic reform and foreign policy matters other than Vietnam. This chapter focuses attention on these other issues and explains what kinds of intellectual were interested in what kinds of issue.

We had best define what is meant by culture crisis, for it is the least clear issue though the most popular one among intellectuals — chosen by almost 40 percent. During the late 1960's the American public at large became increasingly concerned with issues that the news media and the public opinion polls featured as "crime and unrest," or "law and order," "campus unrest," and "polarization and unrest in America." As we saw, none of these issues in themselves were paramount in the view of the public and at least as the polls reported them, they were not, for people in general, merged into a single concern. Not so for intellectuals. The most pervasive issue for them was "the revolutionary character of our times," as one put it, or as another put it, "culture conflict includes everything from the lady with the tennis shoes who feels the world is collapsing, to kids with long hair who want to make it collapse." Law and order was seen as part of this problem of values, but it was defined as a "question of authority . . . the general decline in the acceptance of social institutions as legitimate." Thirty-three of the 108 intellectuals who gave us material on the second issue of most concern to them spoke generally of culture conflict, culture crisis, and values, while a dozen began with the student issues and quickly drifted into these more general concerns. Since it was impossible for most people to talk about one aspect of this complex without mentioning another, we have created this "culture crisis" grouping. This topic is most important for understanding a possible conservative trend among intellectuals and deserves careful analysis.

Race, chosen by more than 25 percent, was the next most popular "second issue" topic among intellectuals. Included in this category were discussions of civil rights, Black Power, integration, and so on. Intellectuals here do not differ from the public in the way they formulate the issues, nor do they, as it turns out, have especially distinctive opinions. Discouragement and befuddlement are common reactions in this area.

The next most often discussed complex of issues among intellectuals, chosen by almost 20 percent of the sample, was economic policy, social reform including welfare, poverty and education, and urban affairs. Grouped under the title "domestic reform," these issues too are difficult for intellectuals to unravel from each other. Both technocrats and Left ideologists are involved in these matters. Both have much to say about these issues and have in fact had some impact on national policies over the last ten years, yet both types feel misunderstood and impotent to effectuate changes they feel America must make in order for it to survive as a just and livable society.

Even after Vietnam had been discussed, over 10 percent of the intellectual elite felt that they had important things to say about foreign policy, mainly in the area of war and peace in general, and the UN, arms reduction and East-West relations in particular. Except that several favored more reasonable relations with China – a step subsequently taken by the Nixon administration – most views expressed were not especially startling or productive. As with race relations and domestic reform, intellectuals in these areas knew what ought to be achieved, have known so for twenty-five years, but admit they have little idea as to how their plans, hopes and ideas can be made into realities.

Finally, four intellectuals spoke up on pollution, population control or ecology, and four discussed a variety of miscellaneous issues such as world travel. The volatility of issues is suggested not only by the small number concerned with environment and ecology, but by the total absence of such issues as women's liberation, homosexual liberation and other similar minority group issues from the list. True, one or two intellectuals mentioned such matters, but only in passing.

This list of issues seems reasonable enough and not especially startling – until we compare the interests of intellectuals with those of other American leaders. Despite the fact that they tend to see the same problems as the rest of the elite, intellectuals are simply in another world when it comes to personal concerns. One year after our interviews with intellectuals we queried other American elite groups. We asked leaders in business, labor, federal government administration, Congress, political parties, voluntary associations and mass media to tell us about the one national issue related to a social problem with which they were most involved. We did not ask separately about Vietnam and Southeast Asia as we had in the study of intellectuals, but some 8 percent of the American leaders picked that issue to talk about. This compares with almost half of the American intellectual elite, who spontaneously selected the topic of Vietnam before we mentioned it. True, Vietnam was a "hotter" issue in 1970 (the year of the invasion of Cambodia), but the Vietnam War and Southeast Asia were still very much with us as national problems in 1971 as well.

Even with Vietnam and Southeast Asia excluded from consideration, as they are in Table 24, it is evident that the concerns of intellectuals are radically different from those of the rest of the American elite. Again, it is true, as we mentioned above, that at the time of our interviews with leaders in 1971, the recession, inflation, and wage-price controls were among the major news stories of the time. Even so, examination of the table shows wide differences between intellectuals and the rest of the American elite, who also differ considerably among themselves. The economy is, after all, most talked about by those who are specialists – businessmen and labor leaders, with the latter feeling especially hard hit. Congress and the administration are next in their concern with the economy, since they are most responsive to the needs of labor and business. It is my judgment, based on the kinds of person in the intellectual sample, that interviews with them timed to coincide with the interviews of the rest of the American leadership, would at most have doubled the number of intellectuals concerned with economic issues, bringing the total to about 16 percent – still a relatively low figure. Since the mass media generally behave like the intellectuals, and since in 1971 only 13 percent chose to talk about the economy, this hypothetical figure of 16 percent for the intellectuals seems quite reasonable.

The most startling difference between intellectuals and other members of the American elite lies in the almost total lack of interest on the part of the latter in the realm of culture and values – the specialty, as it were, of intellectuals. Only 9 percent of the nonintellectual elite chose that topic for discussion. Only mass media leaders, intellectuals' closest kin among the American leadership, were likely to pick the topic – 30 percent chose it. Similarly, except for the voluntary association leadership, some of whose associations were directly concerned with race, the interest of intellectuals in American racial problems is simply not matched by others. The very low concern of the administration[1] and of Congress with race is especially noteworthy. And although ecology as a topic has not yet picked up many devotees, the proportion interested among American

TABLE 24

Issue[a] in Which Leader Is Most Involved, by Institutional Sector (1971) (In Percentages)

SECTOR	ECONOMY	SOCIAL REFORM, SOCIAL PROBLEMS[b]	RACE	FOREIGN POLICY, DEFENSE	ECOLOGY	VALUES, CULTURE	GOVERNMENT, PARTY REFORM	MISCELLANEOUS	TOTAL PERCENTAGE	TOTAL NUMBER
Economic										
Business	40	21	11	2	10	10	4	2	100	129
Labor	58	11	13	2	7	7	0	2	100	45
Political										
Administration	29	19	8	21	14	7	3	...	101	111
Congress	29	24	6	12	14	6	8	2	101	51
Party	16	11	16	5	11	9	32	0	100	44
Voluntary association	18	23	25	11	11	11	0	0	99	44
Symbolic										
Mass media	13	17	13	11	9	30	7	0	100	46
Intellectual (1970)	8	11	26	11	4	37	...	4	101	108

[a] Excluding Vietnam and Southeast Asia.
[b] E.g., poverty, health, education, housing and urban problems.

leaders generally is double that among the intellectuals. Ecology and the environment as issues have of course been cast as technical or scientific matters and so have obtained less interest among the less scientifically oriented persons who make up most of the mass media leaders and the intellectual leadership as we have defined it.

The specialties of the various American leaders contribute to their particular interests in social issues, and it will be evident that specialization even among intellectuals affects the concerns of intellectuals. The very climate created by occupational specialization can lead to a wholly different world view. Concerns with Vietnam, race and the culture crisis give one a radically different view of the United States than do concerns with inflation and the state of the economy as well as with the particulars of health, education, welfare, and other urban problems.

The differences between the American elite as a whole and the intellectual elite in terms of their concerns, despite the fact that they have fairly similar views of what is wrong with America, suggests that what an intellectual sees is wrong does not necessarily predict what he himself will work on. The latter has much to do with his interests, capabilities, and training. Nonetheless, it is important to see the extent to which a person values his own efforts. That is, does he believe that the problems he is working on are of major importance to the country? This is indeed the case with all areas except the culture crisis. One hundred percent of those working in the foreign policy area find it among the most important problems facing the country in the past twenty-five years; 75 percent of those working on domestic reform think it is most important. But despite the general interest of intellectuals in the culture crisis and in values, intellectuals working in that area are no more likely than other intellectuals to see the area as of utmost importance: only 40 percent do so. Is this what is meant by a failure of nerve? Perhaps.

Just as intellectuals differ from the rest of the American leadership and from the general public as well in the topics which interest them, so do they differ among themselves. One of the few generational differences on any matter in our sample of leading intellectuals occurs in the selection of the second issue.

As Table 25* shows, older intellectuals are much more likely to pick foreign policy, whereas younger intellectuals are more likely to choose some topic in the domestic reform area. Although discussants of domestic reform issues tend to be somewhat more radical than others, radicalism in our sample is not noticeably related to age: there are enough older radicals as well as a sufficient number of younger liberals to erase any age differences in political orientation. On the other hand, the timing of issues is very much a generational matter. The issues of nuclear war, disarmament, and East-West relations assumed greatest salience at the end of World War II. The advocates of various positions on these matters in our sample had already formed their opinions at that time. Despite an occasional reference on their part to the interest of youth in preventing nuclear war, the fact is that since the late 1940's, interest in disarmament as an issue has been declining among intellectuals. (See above, Table 23, for trends in issues since World War II.) Fewer of the young people interested in peace now talk about foreign policy, disarmament or world organization. Rather, their talk is about American economic imperialism and military adventurism as a result of the American capitalistic system, and hence their comments are classified under the heading of domestic reform. And as for domestic reform, this too seems an issue of the early sixties, the time when the intellectual interests of many of the younger persons in our sample were formed. Even some of the more conservative intellectuals interested in reform (and there are a fair number of them) formed their interests at that time.

We mentioned that intellectuals interested in domestic reform were somewhat more radical; for example, they tended to be opposed to the Cold War in the late fifties. Intellectuals interested in race, on the other hand, had a more conservative orientation in the fifties, having been favorable to the Cold War. Some of these are the intellectuals we mentioned as having been "radicalized" to some extent by the sixties, though their radicalization extends mainly to the "style" issues of the sixties, such as race, drugs, the counterculture, and so on, rather than to issues of socialism and capitalism.

* Tables 25–30 will be found at the end of this Chapter.

This does not mean that radical circles were exclusively interested in domestic reform, however. When turned around the other way, to see what issues different circles were interested in rather than looking at the politics of those espousing different issues, Table 26 shows that the interests of our few radicals were evenly split among domestic reform, race, and the culture crisis. The representatives in our sample of the New York literary intellectual circle, however, seemed almost exclusively concerned with race. The social science–literary circle seemed split in its concern between culture crisis issues and domestic reform issues. The center circle, as one might expect from its being the epitome of intellectuality of a certain kind, was predominantly concerned about the culture crisis, and the issue of greatest concern to noncircle members also was the culture crisis.

Intellectual prestige is of course related to circle membership, for center circle members all have high prestige. But there are many intellectuals of high prestige in other circles, and some are even isolates. So the question of what problems are being addressed by intellectuals of greater prestige is a separate one, to some extent, from ascertaining which problems are fashionable in different circles. There is no question, however, that very high prestige intellectuals were not only more likely to perceive culture and values as problems but to be interested in working in this area. Table 27 shows that 60 percent of intellectuals who received six or more "votes" were interested in the culture crisis. On the other hand, race was clearly the "wrong" issue, since intellectuals of less prestige were much more likely to pick it. The variety of issues (half were ecological) that we place under the "miscellaneous" label were also chosen much more by less prestigeful intellectuals. Foreign policy was somewhat related to prestige, but in a different way: persons of moderate prestige were more likely to pick it. Finally, domestic reform was unrelated to prestige. This analysis of issues as related to intellectual circles and intellectual prestige suggests once more that Harold Rosenberg's oft-quoted phrase describing intellectuals as "a herd of independent minds" very much applies to the types of issues which they feel are important to discuss. This is of course a matter insisted upon for some time by students of the sociology of knowledge, but rarely does one have such a clear-cut demonstration that some topics are "right" whereas others are "wrong." The findings here are sharper than the findings in mere perception of problems discussed in the previous chapter. As will become evident, the rightness or wrongness of a topic affects *how* it is discussed as well as the mere fact that it is taken up at all.

The appropriateness of a given topic for intellectual discussion may or may not relate to the expertise of the intellectuals who actually discuss it. That is, intellectuals with high prestige may be interested in a particular issue but may not themselves be especially important intellectuals in that field. For each major cluster of issues we therefore noted how often those of our sample who picked that issue were mentioned by other members of our sample as having influenced their thinking on the issue, as having been a discussion partner on the issue, or as having a general reputation for influence on that issue. The intellectuals in our sample were quite important within the intellectual community on most of the issues they chose to talk about to us. On issues of foreign policy, domestic reform and the culture crisis, we have tapped, apparently, an inbred circle of persons who influence each other, for about half those talking about these issues were mentioned at least once as having influenced a respondent's thinking on that issue. Very few of the intellectuals in our sample were important on matters of race, and none on the ecological or other miscellaneous issues. Not only was there inbred influence on opinion, but even a considerable amount of actual discussion within the circles of those who chose particular issues. About 40 percent of those choosing the topic of domestic reform actually discussed problems in that area with at least one other person in the sample who had picked that topic. The same was true for the culture crisis issue. There were fewer discussants in our sample in the foreign relations area than there were intellectual influentials mainly because nonintellectuals who were men of power were frequently mentioned as discussion partners. On the matter of general reputation for influence among intellectuals on issues of foreign policy, domestic reform, race, the culture crisis and ecology, only for domestic reform were half of our respondents mentioned at least once. This again reflects a tendency of the late sixties in which radicals are more visible than nonradicals. Since those picking the domestic reform issue tended to be somewhat more radical than the others, they also tended, because of the nature of the times, to be more visible. (See Table 28.)

Perhaps the most important determinate of an intellectual's interest in a topic is his occupation – the way he earns his living. We defined intellectuals as generalists – persons talking in ways accessible to well educated "laymen" on matters of general significance not necessarily within their occupational specialty. To a large degree this is characteristic of persons within our sample. Nonetheless, when asked, "Which *one* issue are you most concerned with?" half the intellectuals picked an issue that lay within their field of specialization, though the proportion of specialists varied from topic to topic. The occupational division we have been reporting – nonacademic, humanities professor or social science profession – is not fine enough for present purposes (though the data are shown in Table 29). For example, we considered a nonacademic a "foreign policy expert" if he had done much of his work as a foreign correspondent; similarly, a nonacademic who had written much about race would be considered an "expert" on race. Obviously, a professor of government who specialized in foreign policy was an expert on foreign policy, and professors of economics were experts on the economy. But any social science professor was considered an "expert" on social reform, and any professional critic was considered a specialist in the field of culture and values.

Counting specialties in this way, we found that half those talking about foreign policy problems were specialists, two thirds of those dealing with social reform were specialists, and about half of those dealing with ecology and miscellaneous other problems were specialists in one of several fields. The culture crisis was for this purpose divided into two sets of issues: those dealing specifically with youth and student unrest and those dealing more generally with culture and values. Less than half of those picking culture and values in general terms were specialists by any stretch of the term. On the other hand, of the 11 intellectuals who discussed student matters, all were professors. Four of these were research specialists in the field of universities, education and university students (one historian and three sociologists), and their personal research had been in this area for a number of years. But since campus disturbances had affected all of the professors directly and personally, all the professors may be considered "experts" on the campus crisis.

The race relations field is of special interest because of the large number of "amateurs" who picked the topic. Only one fourth of those concerned with race were specialists in this area. A few were historians who had written about various topics on race, and the rest were journalists who had done special pieces on race – almost all of them during the high point of the civil rights movement in the early sixties. Although social scientists had written much professionally about race, and although race relations is a featured course in undergraduate curricula of most psychology, anthropology, and sociology departments, only one social scientist in our sample was among those concerned about race, and he, curiously, was not himself a specialist in this field. Five other academics were concerned with race, and all but one were in the humanities. All the other intellectuals concerned with race were nonacademics, many of them free-lancers or journalists. All of this of course is tied in with the predominant interest of the literary circle in race. But the nonexpert character of intellectuals in this field has important consequences for the way race was discussed in 1970.

We have seen that the social location of a member of the elite strongly affects his interest in social issues. Not only were intellectuals different from other members of the elite, but within the group of intellectuals there were important differences in interest occasioned by generation, political point of view, social circle and occupational specialty. Once a topic is chosen and discussed, it has its own dynamic, however. Table 30 shows considerable differences among issues in the way intellectuals talked about them. An important distinction is between issues which are generally seen as isolated problems and those which seem to be inherently linked to other issues or which are a part of a larger problem. Eight of the 12 intellectuals who talked about foreign policy saw the issues they talked about as quite specific, whereas two thirds of those talking about domestic reform or the culture crisis felt they could not do so without invoking other issues. Interestingly, only a minority of those discussing race brought in other problems or saw racial problems as part of a larger issue.

Then there is the alleged tendency of intellectuals always to invoke a theory as an explanation for whatever position they might take on any matter. This stereotype was not borne out in our conversations with leading intellectuals. Most, when they talked about social problems, did so in a fairly commonsensical manner, eschewing deductions from economic, social, political, literary or cultural theory. This may reflect the fact that our sample was chosen for their having written for general audiences, rather than for specialized ones. Even so, the tendency to invoke a theory varied from topic to topic. No one talking about foreign policy did so from a theoretical point of view, but about one third who spoke about domestic reform or the culture crisis did so. The theories invoked for domestic reform issues were mainly Marxian, while those in the realm of culture tended to be more general anthropological or social theories. One technical mark of the intellectual – a sense of history – was fairly prevalent. Some kind of historical reference was made in the course of discussing the second issue by about half the intellectuals, with the culture crisis, naturally, drawing the most historical references.

Finally, an important aspect of style in dealing with a problem is the degree to which one merely descriptively calls attention to a problem or, in addition to finger pointing, one also calls for action or some kind of change. (See Table 30.) In the next chapter we shall take up in detail the various relations between intellectuals and men of power. At present we wish simply to note that the issues were handled very differently from each other: it is obvious that very few will have any action program for dealing with the culture crisis, but what is especially surprising is that practically no one among those discussing race had any suggestions for action. Even in the realm of foreign policy or among the set of miscellaneous issues only a minority suggested action or change. Only in the realm of domestic reform did the majority offer programs of action. Clearly, most intellectuals do not see their function in social commentary as encompassing specific suggestions for action – it is sufficient for most merely to describe or delineate the problem. And it is these analyses to which we now turn.

TABLE 25

Issues of Most Concern to Intellectuals, According to Age (In Percentages)

ISSUE	AGE		
	UNDER 50 (N = 35)	50–59 (N = 33)	60 OR MORE (N = 30)
Foreign policy	5	8	23
Domestic reform	32	22	3
Race	24	28	20
Culture crisis	34	36	43
Miscellaneous	5	6	10

TABLE 26

Issues of Most Concern to Intellectuals, According to Intellectual Circle (In Percentages)

ISSUE	CIRCLE				
	RADICAL (N = 6)	LITERARY (N = 10)	SOCIAL SCIENCE–LITERARY (N = 35)	CENTER (N = 10)	NONE, OTHER (N = 47)
Foreign policy	0	10	9	10	15
Domestic reform	33	10	37	0	11
Race	33	70	17	20	23
Culture crisis	33	0	34	60	43
Miscellaneous	0	10	3	10	0

TABLE 27

Issues of Most Concern to Intellectuals, According to Intellectual Prestige (In Percentages)

	NUMBER OF MENTIONS, ALL QUESTIONS BUT VIETNAM		
ISSUE	0–1 (N = 52)	2–5 (N = 27)	6 OR MORE (N = 29)
Foreign policy	10	19	7
Domestic reform	18	22	21
Race	35	26	10
Culture crisis	27	26	62
Miscellaneous	10	7	0

TABLE 28

Importance of Intellectuals Most Concerned with Various Issues (In Percentages)

	ISSUE				
IMPORTANCE	FOREIGN POLICY (N = 12)	DOMESTIC REFORM (N = 21)	RACE (N = 28)	CULTURE CRISIS (N = 40)	MISCELLANEOUS (N = 7)
Mentioned one or more times as:					
Influencing respondent's own thinking on this issue	50	57	18	43	0
Discussion partner on this issue	17	43	7	38	0
Having reputation influence on this issue	33	48	11	28	14

TABLE 29

Intellectuals Most Concerned with Various Issues, According to Occupation (In Percentages)

ISSUE	OCCUPATION		
	NONACADEMIC (N = 52)	HUMANITIES PROFESSOR (N = 73)	SOCIAL SCIENCE, OTHER PROFESSOR (N = 23)
Foreign policy	11	9	16
Domestic reform	13	17	40
Race	36	17	8
Culture crisis:			
Values	30	35	10
Youth and students	0	17	26
Miscellaneous	11	4	0

TABLE 30

Intellectuals' Approach to Issue (In Percentages)

APPROACH	ISSUE				
	FOREIGN POLICY (N = 12)	DOMESTIC REFORM (N = 21)	RACE (N = 28)	CULTURE CRISIS (N = 40)	MISCEL-LANEOUS (N = 7)
See issue as part of larger picture, linked to other issues	33	67	39	63	57
Invoke a theory	0	33	18	30	14
Give some history	42	33	39	60	0
Call for action involving some change	42	57	11	23	43

11

The Classic Issues: Foreign Policy and Domestic Reform

During World War II there were two main topics of social analysis and speculation among American intellectuals: what would and should be the postwar foreign policies of the United States now that it was clearly a world leader, and how would and should the United States reconstitute the domestic policies of the New Deal after the inevitable postwar depression. The first topic was taken up by the new internationalists and the coalition of antifascists that was once more reorganized after Hitler invaded the USSR. As the Cold War rapidly became a reality, there remained a group of intellectuals committed to world peace, and these are the ones by and large who chose in our interviews to talk about foreign policy as their second issue even after they had discussed the war in Vietnam. But as we shall see, few predicted or advocated the present turns in foreign policy. For the most part, they had long since had their say and were now all played out.

The area of domestic reform is just the opposite. The rise of McCarthyism, the Cold War, and, to the intellectuals, the surprising success of the American postwar economy, stifled serious discussion of basic domestic reform for a number of years until the late fifties when, at least among some intellectuals, the Cold War began to ease up. Just as the discussion of domestic reform arose anew, the war in Vietnam began to absorb all moral energies, and a conservative administration took office. The result was that while discussants of domestic reform were therefore (because it was a "new" issue) younger than the other intellectuals, their thinking hardly had time to jell, to pick up adherents, to react to new programs and ideas before it was all over. The much heralded failure of social reform of the sixties was not a case of having tried and failed, but of never having had the time to get off the ground either in theory or practice. For there was much new to absorb since the last time domestic policy had seriously been discussed in the late thirties and early forties. Social science had vastly changed since that period and many intellectuals thought that technologies and theories for producing significant social change or reform were now available in ways they had never been before. The issues and problems of social policy are back with us again in the seventies – I believe with a vengeance. But at the time of our interviews in 1970 it was obvious that nothing was clear; the scattered few who were talking about these problems were for the most part confused and pessimistic. So the common thread of issues discussed in this chapter consists of traditional American issues, and for different reasons, those who chose to talk about them had not very much new to say.

Foreign Policy

The first set of classic issues we shall describe are related to foreign policy and war and peace issues. Since we have already devoted much attention to foreign policy we shall be quite brief here. Despite the opportunity to talk about foreign affairs given by our concentration on Vietnam, more than a dozen intellectuals, half of them well-known as professional experts in foreign affairs, felt that they had not yet exhausted their interest in the field and chose the area as their second topic of discussion. And their impact may be greater than their number since four of them were important editors whose journals or newspapers have consistently featured wide coverage of foreign affairs. Moreover, as compared to intellectuals in other fields, those interested in foreign affairs and matters of war and peace have been consistently more politically involved and more active. The UN and disarmament and peace were the main interests of the majority; a second group was concerned with East-West relations and another group consisted of area specialists each discussing his own things.

The UN peace-disarmament group was older than the others and had been active in the field since even before the end of World War II. One social scientist explained:

> I've been working on the nuclear war [as the problem of greatest concern to me] and I started interviewing [on it] two hours after Hiroshima and I've been working on it ever since. Of course it's the number one issue. Because if we are not here it won't make any difference whether we have good inner cities or not.

The issue of sheer survival seems especially compelling to them, for "knowing that the survival of the species is threatened," they "can't understand why governments and political people or just ordinary citizens don't devote more of their energy in trying to work our way out of this predicament." While one expert felt that "if the present trend [toward building MIRV systems] continues, I think that chances are very good that we are going to blow each other up," most felt cautiously optimistic about SALT, the banning of nuclear testing, and the lessened likelihood that an accidental bombing of a city could touch off a nuclear war.

None who discussed the UN or disarmament had any new proposals. One had been a founder of the World Federalists in 1947 and still touted that line and one admitted, "I don't think I have anything novel to say." Rather, their tone was one of the necessity of constant political battle and vigilance. Thus, although moral fervor or the ideology of peace was sometimes interjected, almost all spoke in scientific or functional terms.

Among the small group of intellectuals specifically interested in East-West relations, however, was one editor who spoke sharply and bitterly about American responsibility for the Cold

War and the bad effects of domestic Communist baiting. Interestingly, he tied his discussion of aggressive foreign policy to an analysis of the domestic problem of race. This most crucial problem could have been solved, in his view, right after World War II, had not red-baiting and the Cold War diverted attention from domestic issues. But the other East-West experts were more conventional, albeit in a way that has made them into prophets. Both experts strongly advocated in 1970 the types of detente policy toward the Soviet Union and China which Nixon and Kissinger in 1972 began to pursue. Despite the fact that both are Democrats, at least one has had a strong effect on the State Department and probably also on Kissinger.

A good proportion of the intellectuals interested in foreign affairs had a particular geographic area in which they specialized or which was of special concern to them. Two were interested in the Mideast. A Jewish professor strongly favored Israel, although he disapproved of the Israeli government's relations or lack of them with the Palestinians. A non-Jew formerly in the Defense Department thought that "a *modus vivendi* . . . is going to require the moderation of the Israeli objectives." A Latin American expert criticized our support of dictatorships, while an expert on Germany thought we had better do something about our occupation of Germany after twenty-five years. Considerably younger than the disarmament group and certainly all critical in various ways of American policies, none of these area experts had anything especially new or exciting to propose.

Domestic Reform

The attempts of intellectuals to deal directly with the problems of industrial and postindustrial society are by and large more focused than their commentary about the state of our culture generally or the problem of race relations in the United States, which we shall discuss in the following chapters. Grouped together under the heading "domestic reform" are concerns with the nature and state of the American economy or economic system, concerns with social reform of education, social welfare accounting and the like; and more general concerns with the ills and nature of urbanism. While these concerns can be separated, and most respondents discussed a particular issue, most also recognized the interrelation of all domestic reform issues. Conventions of language, however, force us to talk about one thing at a time, and we shall begin with the economy.

Intellectuals interested in this area are, as we said, on the average younger than those in any other area. Four are academic professional economists, one is a labor editor, another a labor union "intellectual," one is a Marxist historian and two are populist journalist "muckrakers," and one is in a field on the borders of humanities and social science. At least six of the nine are professionals in this area, and the activities of the journalists are such that they too might be considered specialists on the state of the economy.

Despite their relative youth, only several of the intellectuals in these areas were radicalized by events of the sixties and by the change in the intellectual climate which seem characteristic of that decade as opposed to the fifties. Two became interested in these issues in their childhood, though for opposite reasons. One intellectual, the scion of a wealthy family, identified with his family's chauffeur and resented his being ordered about. This, he claims, began his interest in social and economic equality. Another came from a family involved in the New York Jewish radical labor movement who contributed to his Marxist education sufficiently for him to have become, by twelve or fourteen years of age, "sick of Bolshevist ideology." Instead he became a "left-wing Marxist-anarchist critic," a position he still holds today. The interests of the other intellectuals in this area were also long standing, if less dramatic in their inception. Some became involved in the Socialist Party or the labor movement when they were in their early twenties; the rest became professionally interested in economics around the same age and this interest carried them naturally to their present concerns.

Eight of the ten were basically critical of the American "system," though they differ in the extent and nature of their critique. Several took the point of view of one kind or another of what has been called democratic socialism. For example:

> One of the things that happened to me, and I think this had been a fairly common experience, as a result of the war in Vietnam, as a result of the racial crisis, the urban crisis, the whole series of overlapping crises one feels a need for some perspective on all these issues that are related to one another that would enable one to see them not as a series of disparate things . . . to be faced one at a time . . . but as something about the nature of American society itself. As a result of which I and a lot of other people became interested in socialism and the theory that is identified with it—Marxism. . . . We feel that one can't discuss the urban crisis for very long without discovering that it's difficult to conceive of solutions that wouldn't drastically alter the balance of power in American society. . . . The interests of very large corporations seem to work against solutions. [But] a very real difficulty is that Socialism has been identified either with totalitarianism or highly centralized and undemocratic regimes and hence doesn't look very attractive to anybody.

This historian is opposed to forming a third Socialist party but would like to work within the Democratic party — "It doesn't make much difference what the party calls itself" — and "work out the details of what socialism means in the U.S. The kind of party I have in mind is the furthest thing removed from the vanguard or part of a conspiratorial, revolutionary sect." Marxism, in his view is important, "But it seems to me that anybody who expects Marxism to provide all the answers is crazy. At best it can just be a starting point for a very general perspective."

The problem of achieving some form of socialism in a country unreceptive to it is a recurring theme among those who find the current economic system unacceptable. One respondent, a labor union intellectual, spent over an hour discussing the problem. Mentioning a large number of concrete issues such as tax restructuring, race relations and housing, where report after report has pinpointed the need for action, he noted that while corporations are destructive of democratic society, in the progressive wing of the Democratic party there are corporation executives active whose support is important in bringing society further along toward solving its problems. A more radical "left-wing Marxist," who favors "democratic control of economic institutions," saw the necessity for a social revolution but did not know how it might come about. Unlike the labor union man, he did not feel that there are any issues which could be worked out through the present political processes.

> This is a very unpolitical country. . . . Marcuse's "One Dimensionality" is an accurate description of it. . . . One doesn't find any focused political objectives expressed in the United States, just discontent, violence, and fear – irrational reactions to problems sensed but not understood.

Not all who opposed the basic economic system of this country were Marxist democrats. A few were the new Ralph Nader type of muckraker who bitterly exposed corporate bigness – "The development of a country where corporations really run it and which more and more decide public policies." The problem as this man saw it was not that corporations do not work, but that they work too well; this disaster for democracy is not understood.

> Radicals get all steamed up about corporations but they don't care. . . . The fact of the matter is that corporations have succeeded in being very workable organizations.

Finally, there were one or two liberal critics of the economy, on the order of *The New Industrial State*, who noted that the neoclassic model of the economy does not work and that major political power resides with the corporations. They were more hopeful of working within the system of the major parties to reform America than were the other critics.

Included among intellectuals who discussed economic issues were two who took a purely technical view, a view much closer in tone and substance to that taken by members of other elite sectors who discussed economic issues in the larger study of which this is a part. The labor editor pointed out that collective bargaining was failing and was leading to inflation. A younger economic theorist was also concerned about inflation and criticized the way the Nixon administration was handling this problem. In fact, he was the only one in this sample who discussed the economy in a way similar to that of the rest of the American leadership.

The relatively few intellectuals who advocated social reforms of one type or another (exclusive of those specifically wishing to change the economy) were somewhat like those advocating economic reforms: they were relatively younger than the rest of the elite and all but one were professional social scientists with specialized concerns in this field. The two exceptions have for so long been identified with their field as to make them specialists despite their catholic interests and their free-lance status. Most in fact have had a long-term concern with their topic, an interest stemming from their professional activities. One curious exception may prove the rule for experts in social reform proposals. An economist well-known for his monetary theories was asked several years before our interview to contribute his talents to a symposium on a social problem. The economist reported to us that he had never seriously thought about the problem before, but that his solution, for which he subsequently became famous, was one which any well-trained economist would have come up with had he, like himself, approached the problem fresh. While this comment may seem overly modest, the fact is that the proposal was felt by many to be fairly radical, but the economist himself is more accurately defined as a middle-of-the-road liberal. From a technical economic point of view, his proposal is quite conservative. His technical craft, he insisted, indeed led him to the proposal. As this episode suggests, intellectuals proposing social reform were more likely than those advocating basic changes in the economy to have couched their proposals in technical or pragmatic terms.

Aside from the liberal political scientist who hoped for an improvement in the "quality of life" through "tax reform, putting more money into the cities, fighting poverty, and improving services of all kinds," all who spoke about social reform had thought extensively on the topic of their choice and had written much on it. One socialist recounted his intellectual and political history by talking about his twenty-year interest in social equality. Curiously, the ins and outs of the New York so-called

"Trotskyites" received more attention in his account than the specific issues of income and racial equality. Another ex-socialist gave his rationale for working for the last ten or fifteen years on the problem of social indicators, arguing that current economic indicators, for example, do not reflect the true costs of production.

Two of the intellectuals interviewed had important roles in the "War on Poverty." One discussed from the economic point of view his technical solutions for improving the welfare program. There was a sociologist who characterized himself as "strongly conservative" but who agreed with a radical freelancer (even self-consciously citing him) that contemporary schools were showing "more and more signs of wearing down, of atrophy." The sociologist agreed not only with the radical's diagnosis, but advocated part of the radical solution as well: "abolition of the present mandatory school age for children," and replacing it with "a revival of something approximating apprenticeship." The radical, of course, favored a more free-floating experience than that advocated by the conservative, who hoped that corporations would help absorb students. But the radical was also more overtly ideological in his tone, saying, for example, "I think the elementary education ought to try to keep the kids wild rather than socialized. . . . As for the colleges, the only useful higher education is for professional training. The liberal arts college is nonsense."

The other intellectual concerned with poverty was more ideological and less purely pragmatic.

> My program includes so much change that nothing short of a major political movement is really going to bring it about. I think it's nice for the educated and concerned people to get concerned about this or that issue, but unless there's real political change people are just going to stumble from fascism to poverty to the war to ecology and never really solve any of them — just generate a lot of talk.

On the other hand, in discussing various social reforms he noted, "I think structural reform is necessary but you can do this on the basis of liberalism. You don't have to be an anti-capitalist to do it, although it's quite a bit easier." This pragmatic quality is reflected in his proposed strategy for abolishing poverty.

I think we made a mistake. We tied the anti-poverty program politically and programmatically to a minority of perhaps 20 per cent of the American people. This, in effect means you've got to convince the other 80 per cent to do something for the 20 per cent. I would now claim the program as a poverty emphasis within a broader program that would provide housing for all Americans, including tradesmen who can't make ends meet on $8,400 a year after 21 years of work. That is, to try to hook the poverty issue more into the working class issue.

Finally, several literary critics and one urban specialist were concerned about urban problems. As might be expected, the urban specialist took a fairly technical view of urban problems, but the critics were mainly concerned about the quality of urban life. All were liberals rather than leftists. Perhaps in talking about urban problems they avoided direct attacks on the nature of our system as a whole. One Victorian-era scholar expressed only a general concern and interest in "the problem of urbanization and what connection in an evolutionary way there is between the urban experience in the first part of the industrial revolution and how it changed and how certain continuities persisted to the present." Though himself a city-dweller born in New York, his adult interest had been sparked twenty years ago by work on Dickens, Balzac and other nineteenth-century literary figures who wrote about the new urban conditions. Another Victorian-era expert, who specializes in American rather than English literature, compared the New York City of Dreiser's day to the New York City of today, and found that "for every class living in the city has become more painful." And the editor of a literary review complained about

. . . the physical deterioration of the cities, which I find absolutely appalling. The measures that have been taken to correct these conditions have been in so many instances aesthetic and social disasters of which I have quietly despaired.

The social scientist had made a number of specific proposals to improve city living, while the literary figures were more inclined, as they said, toward outrage and despair, and though all were professionally involved through their writing and teaching in urban studies, they were less clear than the social scientists as to what might be done to improve life in the city.

The general character of the discussion on social reform is more vigorous than most of the discussion about urban affairs, and more informed than discussions specifically centered about race. A number of concrete suggestions were made and many intellectuals have had some hand in actually trying to work some of them out. On the other hand, the more radical intellectuals, and they are a majority in this field, feel they know what ought to be done but do not know how to implement the massive changes in American society they feel must take place for the world to become more just. But if these discussions of foreign policy and domestic reforms seem inconclusive, the lack of closure is even greater on matters of race and culture, which will be taken up in the following chapters.

12

Race Relations: The Expulsion of the Intellectuals

Most of the intellectuals we interviewed were followers, not leaders on issues in the race relations field. Their interest in race relations generally followed the surge of civil rights activism of the late fifties and early sixties. As we saw, in the five-year period 1945–1950, only 15 percent listed race relations as the problem of first or second most concern to them. In the five-year period 1950–1955 only 22 percent listed race relations as the top or next to the top social or political problem they were personally concerned with as intellectuals. But when the Civil Rights movement picked up momentum in the five-year period 1955–1960, over 40 percent gave race relations first or second rank, while in the next five-year period, 1960–1965, more than half so ranked it. In 1970, almost 30 percent were willing to talk about race relations as the one issue they were personally most concerned with. In many respects, the intellectual elite were not much different from the population at large, since Gallup Polls show that in the 1960–1965 period more than half the country thought race relations was the "most important problem" facing

Stella Manne wrote a first version of this chapter and contributed both analysis and conceptualization.

the country. Especially in the period after World War II and right up to the surge of interest in civil rights, elite intellectuals concerned with race tended, as we noted earlier, to be less prestigious than other elite intellectuals in our sample. We should also remind readers that most of the intellectuals choosing this topic were "amateurs" rather than experts on race.

There are further anomalies in this field. We interviewed only white intellectuals. Few Blacks were on our initial list garnered from the journals and all refused an interview. Had we "snowballed" on the basis of the issue of greatest concern to intellectuals separately from all the other questions, we would have included more Blacks. But the snowball interviews were collected on the basis of all the sociometric questions and so tended to be heavily weighted in the direction of Vietnam experts. The intellectuals whom we did interview who were interested and active in the Civil Rights movement had just recently been pushed to the sidelines by the new line of Black militancy. As one respondent told us,

> When I was active in the Friends of SNCC four or five years ago, the issue seemed wonderfully clear. We supported those organizations – all of them were both Black and white, though perhaps predominantly Black, as was SNCC – which were working by non-violent means for an extension of Negro rights all the way down the line, particularly in voting, but also in housing and the rest of it. . . . At least for a while, SNCC and the Southern Christian Leadership Conference did seem to be accomplishing a great deal. . . . We could use money raising affairs not only to raise money but also as propaganda because both the newspapers and the TV stations in this area were on the whole pretty favorable to that degree of Negro activity. Then, of course, when Stokely Carmichael made that speech for Black Power, they just threw the whole thing down the drain.

The war in Vietnam sapped further energies from the Civil Rights movement. Those intellectuals who received a large number of nominations as influentials on the war were among the least active or interested in race issues. The war in Vietnam, the urban riots of a few years before our interview (riots predicted in the published work of some of our respondents), the lack of forward movement of the American people on race issues and their own exile from the "movement" have led most of the intellectuals we interviewed to a pessimistic view of American race relations.

One of the more important determinants of orientation in the field of race relations, aside from circle membership, was the way an intellectual became interested or involved in the topic. Several said their interest stemmed from their early childhood, at which time they became conscious of race differences – usually in a painful way. Entry into the adult world was another important source of interest in race. For some this meant World War II, when the facts of racial discrimination were dramatized in brutal ways to Northern liberal white intellectuals who befriended Black soldiers. These liberals had never before shared their daily life with Blacks and were brought up sharp by discrimination both within and outside of the military. Others became aware of racial problems in their first jobs. Together, childhood and first adult experiences account for about one third of those now interested in race. Professional interests accounted for another third of those now concerned with race. These included professional academic interests in Black history, for example, as well as the dramatic experiences of reporters who covered the early Congressional debates on Fair Employment in the 1940's or the more dramatic civil rights sit-ins and marches of the 1950's and early 1960's. The journalists who covered the active civil rights period knew almost all the major civil rights leaders, and had walked together with them on many marches. Finally, one third of the intellectuals who were especially concerned with race had become involved in the Civil Rights movement as a result of events experienced at second hand – the reports of the early battles over school integration, the marches, the sit-ins, and the urban riots. The Supreme Court decision of 1954 had great impact even on intellectuals. One might have imagined that intellectuals were those who forced that issue, and indeed some were. But many in our sample were stimulated by the official stand of the court, showing that intellectuals as well as other citizens are affected by the value-climate created by the national government.

Most of the intellectuals concerned with race had some personal experiences in race relations and confrontations; most knew one or more well-known Black civil rights leaders. We asked systematically, as we did in other fields, what books, authors, or persons they knew had influenced them in their ideas on race relations or whom they considered important in the field. Not everyone was mentioned as a person agreed with, merely someone whose ideas they had to take into account. The list, constructed from the nominations of twenty-three respondents, is very instructive.

Top Influentials on Race

NAME	NUMBER OF NOMINATIONS
Malcolm X	11
James Baldwin	10
Eldridge Cleaver	9
Daniel Patrick Moynihan	8
Martin Luther King	8
Kenneth Clark	7
Gunnar Myrdal	6
Ralph Ellison	5
Richard Wright	5
Bayard Rustin	4
C. Vann Woodward	4

There are only three whites among the eleven on the list. And reflecting the literary background of intellectuals interested in race, there are only three social scientists. As one editorial writer remarked,

> In the whole field of the sociology of race there's an enormous literature which I have read, some of which I know exists, but I am not conversant enough with the field to conjure up names.

The word on race came primarily from Black novelists, essayists and politically active civil rights leaders. Again and again respondents emphasized the understanding and feeling that came from reading essays and novels, and above all, Malcolm X's autobiography. Because our intellectuals were all white with necessarily limited experience of Black existence, the humanistic writings gave them a sense of immediacy they could understand and identify with. Typical are the following comments from a literary critic.

> I have been influenced by Ralph Ellison and Jimmy Baldwin in ways that others have been influenced by certain writers that they liked. . . . My friend Ralph Ellison . . . is a great and wonderful writer and I think his new book is the best novel written by a Negro. But I knew Richard Wright in the '30's and I've known several of the [Negro] writers and I'm very sympathetic to writers. They have a tremendous human and artistic quality which I think as a writer and a critic all do good for me. . . . To me, Richard Wright, James Baldwin, Ralph Ellison, certain poets, have been the best writers. . . . Now one of the books that has influenced me very much recently has been the *Autobiography* of Malcolm X. I love it, I think it's great. I am not an admirer of Eldridge Cleaver, I should explain, I am not an admirer of Negro militancy. I am not an admirer of Le Roi Jones. What I admire about Malcolm is his incredible scriptural sense, . . . whereas Cleaver seems to be just another militant. . . .

Despite the negative comments about militants and Black separatists, this critic did take the militants into account. The top three influentials on the overall list were, of course, all militant.

Moynihan's position is another matter. At the time of the interviews he was Nixon's adviser on domestic policy and had recently come out with his doctrine of "benign neglect." Respondents conceded he was important when it came to policy on race, but the following comment, perhaps playing on the word *benign*, was typical of the more kindly remarks:

> I can never tell whether Pat is benign. I guess he's been a good influence. It's awfully hard to tell. Nobody's right and nobody's wrong anymore.

Most of the journalists said they had read one thing or another on race, or had long personal conversations with some of the civil rights leaders, but in general they felt that events and the Civil Rights movement itself had more to do with the climate of opinion on race than any work by intellectuals.

I think it was an idea whose time had come. I think intellectuals were peripheral to it. . . . The Kennedys – Hell, they didn't begin it, they had to run to catch up with it. I think everybody did. It was just something that happened. It was the right time and the right place. . . . I don't think intellectuals were ever knowledgeable about it. I can't really remember reading in the late '50's or early '60's very much intelligent reporting or writing about race.

By 1970, what the idea was, or what even the problem was seemed in dispute among those intellectuals who especially cared about race. About one third of the respondents felt that on the whole, racial attitudes or psychology were the root of the American race problem, although some taking this position admitted that political or social structural factors were important. Typical of those taking a psychological view was this respondent, who had a long-term interest in race problems:

The basic problem is that the overwhelming majority of white Americans are committed, either consciously or unconsciously to maintaining a white dominated society. . . . I think these are unconscious attitudes and everybody sort of accepts these attitudes and moves to their positions on the basis of these assumptions.

As will become evident, a position emphasizing the innate psychological character of race relations is hardly conducive to optimism. "Unlike the War in Vietnam [the race problem] depends on what people think. . . . Both myself and the whole country throughout its history being naively and blandly racist." Not surprising, in view of our white sample, is the overwhelming tendency of respondents to interpret racism as the white man's Negro problem rather than the Negro's Negro problem. Psychological obstacles faced by Blacks – apathy, self-hatred, low self-esteem – which had been discussed widely in the past (albeit more in the social science literature than in the novelistic one) were mentioned by only one respondent, an editor who proclaimed a number of times during the interview that he was not an expert on the problem and knew relatively little about it.

Perhaps the lack of interest in Black psychology is occasioned by the very growth of overt Black expressions of anger. A free-lance journalist who had once written about the damage that 350 years in America had done to the Negro's personality now focused his attention on political and economic factors underlying the race problem. Things had changed, he felt, since he wrote his book.

They've come a great deal further and the lid is off their anger, whereas, in the past, Black anger was still under considerable control. A big piece of the Black man's Negro problem was that the anger had to be suppressed or he couldn't survive, and so the anger turned inward to self hatred or hatred of other Blacks. . . . There is an enormous liberated feeling from the discovery that after 350 years of fearing whites, it's the other way around.

Once psychology is dismissed, though almost all intellectuals feel it is at least an aspect of the race problem, then social class and power become important factors. One third of the intellectuals thought structural matters were at the heart of the race problem. An editor explained:

You really have created – in a society that doesn't believe in a depressed class, doesn't believe in its necessity and rightness – a kind of permanently depressed class.

Depression or deprivation need not be simple poverty, as a syndicated columnist noted, "Poverty is not the problem – prosperity aggravates the lack of mobility of uneducated Blacks." Sheer lack of power may be the underlying problem, according to a literary critic.

Since World War II, America has become a superpower, sometimes the "Big Brother" mentality. . . . The Negro-white issue is . . . in many ways more fundamental than Vietnam because the Negro is special to America. And to me, the name of the game is power. . . . I am more and more struck, living in New York . . . [by] its millions of insignificant people living in Brooklyn, Queens, and the Bronx, its great masses of disregarded and neglected Puerto Ricans and

Negroes, and at the same time, by the extraordinary power of the United States.

Finally, another third of those concerned with race focused more on political problems than on sheer structural ones, though of course the arguments are related and many respondents mentioned both politics and social structure in the same breath. Typical of the more political statements was this complaint by a political activist:

> The Nixon administration doesn't seem to care about civil rights. . . . A lot of the necessary legislation is there; it simply has to be talked about and upheld by the President and the federal government. They have to keep telling the people.

The three orientations toward the race problem – psychological, structural, and political – are obviously not competing views; most concerned intellectuals would, if pressed, admit that all three are important aspects of the problem. They do represent differences in strategy, however, and the fact that concerned intellectuals were split in their emphasis is a sign of the confusion of the times. To be sure, the "line" intellectuals take today is not chosen at random. The strongest determinate is the way an intellectual first became interested in the race problem. If childhood or early adult experiences were crucial, then he was almost sure to emphasize the psychological character of race, for the shocking experience of his own or others' racism still remained the most important fact in his understanding of race. On the other hand, two thirds of those who came upon the field through reportorial assignments or other professional interests saw the problem as a structural one. Structure is almost the universal "line" of published intellectuals with a social science bent, and it seems to have prevailed over other views for most journalists who have covered race relations. Perhaps it is the most convenient view for the detached intellectual.

The Civil Rights movement converts were split between psychological views – "white racism" – and political ones – the failure of the movement to capture the attention of the current administration. Since racism is the view advanced by the more radical wings of the movement who seem nearer to the scene, and political action by the more liberal wing which had been around a long time, the Civil Rights movement intellectuals seem fixated at the point in which they first became involved in race relations. The most important thing about intellectuals and the Civil Rights movement is that the white intellectuals we interviewed have been cut off and divorced from the movement for a number of years.

Whether intellectuals take a psychological, structural, or political view of race relations affects their other opinions on race. While two thirds of the intellectuals concerned with race still endorsed the idea of integration – "I think integration is the only possible ideal for this country" – one third did not, mainly because it is not now feasible. Those maintaining that psychological racism is the main aspect of the race problem in the United States were less likely to approve of integration as a workable goal. The reluctance of the minority of intellectuals fully to endorse integration is also a result of the influence of Black nationalists, whose most powerful influence in the intellectual community has been in the literary, center and radical circles (but definitely *not* in the social science–literary circle). This influence was of course reflected in our list of influential intellectuals on race as seen by our respondents.

Black Power expelled the white intellectuals from the Civil Rights movement. Naturally, a good many respondents had something to say about Black Power, and their points of view generally coincided with their views on integration. But not completely, because many thought the term "Black Power" very unclear. A sociologist expressed the view of the overwhelming majority of those who discussed the matter:

> I'm quite skeptical about Black Power. To begin with, I don't know what the hell it means, and certainly [I am skeptical] about some of the Black Power advocates like Bond, Stokely, and especially, of course, the Panthers. If Black Power means the creation of a kind of Negro moral community with a certain set of values, and so forth, then fine. If it shades into ideas of Black separatism and fantasies about taking over this or that piece of territory, be it Harlem or some place in the South, I think it is paranoid.

Others were more directly "disturbed by the extremists of the Black Power movement, especially by violence." Only a very few supported Black Power as expressed by the Panthers. One radical explained:

> It is important to support the Panthers. Their militarism is a necessary part of the Blacks' own fight. They may seem extreme and radical as compared to the others – King, Wilkins, etc., but as they abandon each of their positions for more radical ones, the others take up the abandoned positions, because they don't seem so radical any more, by comparison. So all groups within the Black community do have a function, though they may not seem to be working together.

Not surprisingly, the few supporters of Black Power were more likely to have a psychological view of race problems, and more importantly, to be members of the literary, center, or radical circles.

We now come to the most characteristic expression of the position of American intellectuals on race: pessimism, or at least, the absence of optimism. Some, especially the advocates of a psychological explanation of race, were pessimists even about the long run, for example: "We're in very bad trouble. . . . I'm more pessimistic now than before." Or the sense of gloom of an "Old Left" radical:

> What I see happening is a combination of polarization and reghettoization. . . . I don't know if we'll get through this period. I think there is going to be an enormous amount of internal turmoil and a growing sense of police state.

Others set some vague time limit to their pessimism, feeling that in the short run things would be bad but "in the long run we're going to luck it out, but it's going to be a slow process." The essence of the response to race problems was a mixture of doubt, despair, confusion, and some wishful thinking. A freelance writer summed it up this way:

> The only question is whether democratic institutions can survive this phase in the evolution of race relations in America. If we can, we

might go on to something better. But I don't know. There is grave doubt.

To be sure, some few had a ray of hope, as did this fairly conservative editorial writer:

> I think we have been going through a bad time but I look upon the last ten years, despite all the upheavals, riots, and so forth, as a time of great progress for Negroes. I think you don't make progress on a difficult issue like race without violence, and so I'm still basically optimistic.

Gloom was most associated with psychological views of the problem, because "human nature" could not be changed. Similarly, members of the literary, center and radical circles who had much invested in the Civil Rights movement and in the growth of Black nationalism now were among the most pessimistic, since their hopes seemed to be shattered by the failure of the movement to bring its message home to white America as well as by its own internal fragmentation.

Leading American intellectuals have been left impotent and confused on race by a combination of two basic factors: the expulsion of white liberal and left intellectuals from the Civil Rights movement together with the development of a vigorous Black intellectual expression — mostly separatist in nature — and the "benign neglect" of the Nixon administration coupled with its growing resistance to the ideals of integration. While seeing the problem as perhaps the most crucial one facing the United States, these American intellectuals find that neither Blacks nor whites will listen to them. Moreover, they once thought that the solution to the problem was known, that it was merely a case of helping the Blacks to become stronger and convincing the white majority of the justice of integration. Very few intellectuals now have a clear idea of how to solve the race problem and do not know what they should now propose. Their reaction is to back off from the issue. As one editor put it to us when he rejected race as an issue to discuss with us:

> Most things we take for granted. We don't sit around, my friends and I, and tell each other how awful it is that Blacks are treated . . .

We don't talk about the Blacks because there's nothing to talk about there. I mean I'm for Blacks being equal. I don't know what there is to talk about!

The problem as most intellectuals see it, however, is not that the problem is simple — they agree that it is, but that the solutions are no longer obvious.

The position of most on race is expressed by an art critic who in the course of a freewheeling interview covering a variety of topics had this to say:

Anyone who is not impelled to vacillate from time to time on this issue [race] is either a fanatic or a moron. This issue is so profound and so vast, involves so many sacrifices of value, that anyone who doesn't realize that this is a fundamentally tragic question can't understand the question. . . . I'm more intelligently confused than I used to be.

13

The Great Culture Crisis

If one set of concerns can be said to be the specialty of intellectuals, then it is surely culture and values. As we have seen, it is not that other elites, or for that matter the public at large, do not see culture and values as a major American problem. They do, and no less than the intellectuals do they rate these matters as being of great importance. But they are either not prepared to work on them themselves or feel incompetent to do so. If there is a crisis in American values, the rest of the elite are clearly going to leave the matter to the intellectuals to solve, if any solutions are possible.

Most of the discussion by intellectuals in the area of culture and values was of broad importance, but there were formulations which were narrowly construed. Some few discussions (six, to be exact) stemmed directly from the intellectual's long term professional concerns. These were interesting and highly influential within specialized fields, but were sufficiently idiosyncratic for us to summarize them here only very briefly before we go on to the central concern of "culture crisis." Some of the "culture specialists" discussed the state of the various arts. The mood was negative. A film critic spoke of censorship in the movies — he was against it, but quite disappointed with the way the new rating system he had originally touted had actually worked out. Another critic who covers many media assessed the overall state of the arts. He found them all deficient, but rated fiction first, then film, with theater at the bottom. A writer talked about the influence of writers upon other writers — he deplored it. And a critic who had long opposed government interference in the arts now conceded that some grants, properly protected, might be useful. Aside from the arts, there were two other specialized concerns. The impact of biology on cultural values was discussed by a biologist, and the failure of the Catholic Church to reform itself was discussed by a Catholic activist journalist. All these concerns hinge in some ways on the general state of American culture — are affected by it or in turn affect it.

On the other hand, concern with the student revolts of the late 1960's seems on the face of it to be a narrow issue. In fact, discussion of the student problem invariably led directly into a discussion of the general "culture crisis."

The main event, of course, was one version or another of the "culture crisis," chosen by four out of five intellectuals who talked about culture. Unlike culture specialists who reflected a long-term interest in their particular subject matter, most of the intellectuals talking about the culture crisis were responding to the recent events of the 1960's. A response is almost by definition a reaction to something originated by someone else. Intellectuals who were concerned with the culture crisis were by and large on the defensive, acting the role of establishment rather than vanguard. Instead of setting up new standards or presenting new ideas they tended to defend the old. As defenders of the status quo, in the strictest sense of the term, most were conservative rather than progressive.

A few saw the culture crisis in terms of the relation between the American value system and the American polity. One way of putting it was that we in the United States had an authority problem. A respondent who picked this topic was then asked by one of our interviewers, who posed as a wide-eyed innocent young lady, "Is there any overriding authority problem in this country?" To which our man replied, "There sure is, Whitey!

Why do you think the streets are full of people beating each other up!" Though his complaint had specific roots, when asked about what could be done about the problem, he was quite abstract, as were most who talked about a culture crisis.

Well, you don't know much about that. Authority relationships are consensual, and that distinguishes them from power relationships which are coercive. When consensual relationships start to erode, people don't know much about how to stop it.

The key event which brought the problem to his attention was the assassination of President Kennedy, "Not so much the assassination, but the response to it which I thought was insufficiently sensitive to the idea that things could become unstuck easier than anybody realized. . . . That we would spiral into a position of increasing sense of non-legitimacy of institutions and the fact that what the government said was not necessarily what the government did, that things could not be believed."

A muckraking journalist, much more to the left of the first respondent, was also impressed by the Kennedy assassination, reaching the very conclusion which made the first so apprehensive. "I found the damn thing falling apart. . . . There had to be two gunmen." This writer's concern with "the right wing" really began, he said, with Roosevelt's reelection in 1936. "They began this hate-Roosevelt stuff. It was really vicious and it was all built around the theme that any liberal was a socialist or a communist." All of this has led him to question "whether we are going to keep a democratic society." The obstacle standing in the way of such a society is "a very tightly held power complex. Your big industrialists cooperate with the military and with the CIA." This muckraker was joined in his feelings by a younger journalist with a similar reputation, who expressed a "general concern with public policy, the nature of the American State," in which "power rests more and more in the hands of the few who are unaccountable." A more liberal respondent wondered about the "effect of large corporations on societal values," noting that "as social changes take place . . . individuals at the bottom of the heap are going to get screwed."

These concerns which explicitly link values to social structure were generally critical of existing structures. But criticalness in this sense was a minority view. The specific structural crises of the University in the late 1960's elicited from the professors – and as we said, all who discussed it were professors – an almost uniform defense of existing values, together with only a grudging admission that structures might have to change. To begin with, almost all the professors who talked about university problems, whatever their position, agreed that the university crisis was the result of, or symptomatic of, a larger crisis in Western thought, values, or social structure. And it is for this reason that we group the university problem together with the problem of culture crisis. Only one intellectual we interviewed was entirely on the side of the students and even he said, "I'm still ambivalent. Sometimes I write out of one side of my ambivalence and sometimes out of the other and sometimes out of both." His analysis of the nature of the student revolt was widely shared, though his prescriptions were not:

> It amounts to a change in the very basic notion of what it means to be human, the end of the Renaissance idea of man. . . . What's mostly involved . . . is the collapse of the notion that growing old ought to mean growing up. . . . It is a problem which wracks communist and socialist countries just as it does so-called capitalist ones. The difference between a cultural revolution, which I think of as the only real revolution (political revolutions barely touch the surface) and a political revolution, is that cultural revolutions are never called for; they're never described in manifestos. Only afterwards can they be appreciated and recorded by analysts.

As he saw it, "Some of the students in part were protesting against what was basic to the university: certain beliefs that say rationality was preferable to passion." While this professor preferred nonrationality or passion, almost all the others were firmly on the side of rationality. But their perception of the students' position was identical to that of their more radical colleague. For example, one professor who felt that "unless we reach a turning point soon the University of California is going to be shut down," also felt that the "course of student revolt is not reform or even what I would call a constructive effort to come to grips with the problem but rather a sort of infantile rage." In fact, every professor who talked about university problems except the first was of the opinion that rationality was a main issue, and that the students opposed rationality. Given the general agreement that the University has historically stood for rationality, that intellectuality means rationality and the perception that the students were nonrational if not antirational, it is not surprising that almost all among the intellectual elite who spoke about student matters were strongly opposed to the students.

Underlying the issue of rationality was one of work ethic. According to our respondents, students do not want to work in the way their professors did. There is a "crisis of vocation. . . . Young people are going to the university because there is nothing else to do . . . not to pick up skills preparatory to earning a living." The one radical professor agrees, but rather likes the idea of play and hedonism.

> If you live as we do in a time of social and cultural change . . . then education involves not history and memory so much as prophecy and vision. . . . Isn't the university, in order to survive, going to have to turn into a place in which the young instruct the old, rather than the old instructing the young. . . . You say, what can the old teach the young? How to grow old. And what can the young teach the old? How to stay young. . . . Can the university turn itself around the past to the future . . . , become interested in teaching people non-specialization . . . become a hedonist rather than a utilitarian kind of institution?

The themes of culture crisis, rationality and vocation were the major issues in analyzing campus unrest as a turn "against the values of Western capitalist liberal democratic society," but others were mentioned as well. One professor, for years vigorously "anti-Stalinist from a democratic-socialist point of view," felt that the New Left was "eroding the amenities of rational discourse."

> Anybody who takes a public position which is contrary to the views of a small radical minority not only suffers the humiliation of being showered with obscenities and abuse, but even runs the risk of physical assault. . . . When we were radical students we never tried to shut anybody up. We were always looking for an argument.

Several professors thought that the students were responding more directly to a failure of the United States to solve racial problems, to end the war in Vietnam and more generally to a failure to apply its vast technological resources to improving its own society. One said, "There is no question as to what they claim to be for: racial justice, more sensitive approaches to the environment, and so on." And another said, "I'm in agreement with them, if not their methods all the time." Those who took the first position, that the university crisis represented a rejection of American and Western values, said that although they had once thought that students were merely affirming the American value system and chastising adults for their failure to live up to their own values, they now felt that the issue was more fundamental. "This is a very basic intellectual-spiritual crisis in a society which hits the universities hard." The professors who took the more global view of student dissatisfaction were those who had devoted much more professional energy to the campus and the student problem than those who felt students had more limited discontents. The kinds of solution professors offered for the campus crisis and their degree of optimism in turn depended on their view of the nature of student rebellion. Typical of those who now feel the problem is a fundamental one is a professor who was at Berkeley during the initial student upheavals, who favored reforms but who now feels that the reforms which he advocated and which were put into practice seem not to have made much difference. "My views are less defined [now] even though I've written more about it." He feels that "much of it [the student position] has merit. . . . Then I have a change of mood and I think that regardless of what's wrong with the institutions of higher education, certain kinds of elements that are under attack – organization, structure, discipline – are essential elements even if in any particular case they are not well suited or justified. . . . So on the whole, this is a revolution that really could not be accommodated within almost any society." Nonetheless, even though the crisis is essentially outside the University, the only thing the University can do is to respond in terms of educational change. Whether this will really help, he is less certain than he was in earlier years. Both the firm advocate of the revolution and the many opponents fear "backlash." As the most radical professor puts it, "The real conflict is not between the community and the university but between parents and children," but some scapegoat must be found and the University is the most likely candidate.

Professors who felt students are upholding the best in Western or American traditions felt that "the answer to this [the student crisis] lies in a meeting of minds which I believe to be possible if people will just take off their masks and quit playing their roles and act human. . . . Those of us who treat every individual as though he was a person, regardless of the age difference, soon find the age difference is no insurmountable barrier." These professors believed that "bringing students to the conference table" would go a long way to alleviate campus problems.

In sum, almost all the professors had serious misgivings about the student revolt. Only one professor strongly favored it, and another more conservative professor thought it could be dealt with. Quoted above as believing in the absence of an age gap, he was the only editor in our study capable of naming a large number of up-and-coming young writers, even though he himself was well past sixty. Thought almost everyone we interviewed favored legalizing marijuana, this professor, sympathetic to the young, openly admitted he would like to try it.

> I have no scientific endorsement for this, but I'm convinced by now that it [marijuana]'s less harmful than what you're smoking [the interviewer was a heavy tobacco smoker], and not as debilitating as liquor in large quantities. I've never had any. . . . Sometime maybe up here [in his study] I could have pot once.

But aside from two essentially unsympathetic social scientists who "understood" the student revolt, all the other professors felt that the students were basically wrong, and some thought them dangerous as well.

The majority of intellectuals interested in culture and values spoke more generally about the quality of life, the nature of Western values, and the crisis engendered by the rise of the counterculture (as distinguished from the adversary culture) in the 1960's. The intellectuals speaking in these general terms were among the most prestigious in our sample, as one might expect, for a general concern with values is the very mark of an intellectual. Of the twenty-two who discussed culture and values in such a broad way, only six favored the counterculture. All the rest were in one way or another opposed. And as one might suspect, the attitude they took toward those under thirty was for the most part highly indicative of their general stand on values. There were basically three positions on values: the liberal position, an ambivalent one, and a more or less pro-counterculture one. Eight of the ten liberals mentioned youth – all negatively. Three of the six who were ambivalent about the counterculture mentioned youth, and those who did were ambivalent about them. Only one of those favoring the counterculture mentioned youth directly and favorably. The rest mentioned them only peripherally. For those favoring the counterculture, youth were simply not an issue.

The liberal position is the dominant one, so let us look carefully at some of its components. It is best to begin with a critic who specifically says her topic is *not* the student revolt.

> Student unrest doesn't interest me. I'm interested in the formation of a Left Wing movement in this country. . . . My primary concern about this is that it is an irrational force in our culture. It has no ideology. It is not thoughtful. It doesn't have programs. It doesn't foresee a society as a result of its activities. It doesn't go down well with an old-time Marxist [like me].

Several recurrent themes already met in more direct discussion of the student issue are repeated: the irrationality of the New Left, and the general disdain for its lack of system by former Marxists.

A more general liberal position is taken by the critic who talked about "cultural conflict. . . . Many of us feel that in addition to living through a social crisis we are living through a crisis of civilization." The sense of collapse was shared by a novelist who said, "We are in a terrible disorder. . . . We are riding fast toward a smash. . . ." As in the case of student disorders, liberals felt that a lack of interest in achievement was both a symptom and a problem in itself. "Western Civilization," the novelist went on, "has become an amusement civilization. The motives for work have worn themselves out." But the importance of play was not the invention of the youth. Intellectuals themselves, or at least some intellectuals, have been responsible.

Margaret Mead introduced barbarian playtime into American life – free self-manipulation and erotic swinging à la Samoa in the South Seas, etc. . . . She and Paul Goodman have probably done more to undermine the foundations of life in America than Marcuse or any other imported "Dutchman."

Finally, several liberals complained that, curiously, disdain for the achievements of Western culture was really an elitist position.

The natural leadership of the country – the upper class and their sons and daughters – have lost confidence in the country and beyond that have lost faith in themselves and in the underlying principles in Western Civilization. . . .

A person who would have been considered a candidate for the elite in past years probably doesn't like his parents, doesn't like the economic system, probably doesn't like his religion very well, and so then he either turns to active radicalism or just drops out and becomes a pretty bad citizen. . . . The reaction to this is very intense from what today is known as the "hardhats" – what was last year called the Blue Collar workers.

Another intellectual found the current "very peculiar alliance [in the politics of culture] between Liberal Republican and Left-Wing thinkers . . . extremely hostile, to put it mildly, to the welfare of people who work for a living, especially proletarians."

In short, the liberals see their world of culture and rationality as crashing down upon them; their students and the younger generation reject the time-honored values. To make matters worse, alienation from high Western culture is not bringing youth closer to proletarian culture – a goal most intellectual leaders once thought desirable, but on the contrary, youthful hedonism alienates the working class.

Then there are the ambivalent liberals. Their diagnosis of the breakdown in culture is identical to those more certain of the apocalypse. But they are less certain that the new culture and the youth identified with it are totally without redeeming features. For example, after he had given a picture of the "quality of life in America – it has to be as big as that," similar to that given by other liberal respondents, we asked this literary critic, "What is your feeling now about this new culture?" He replied:

> There I'm ambivalent. It's not producing a great deal. Many qualities are very praiseworthy. I think young people are sweeter than they ever have been. Their kind of easy responsiveness is quite impressive – the sense of being able to talk to anyone – simple, direct, without being awed or being shy – quite remarkable and quite engaging. At the same time I am concerned with what is lost in personal definition. My notion of art comes from those things that we had called achievement. They come exactly from that often aggressive definition of self.

Others were ambivalent not because there were some redeeming features among the young, but because "the main problem of our time is that it is a revolutionary period," and this in itself is confusing. At one time, this critic explained, the "world had a nice cover of Western Culture. In the past 50 years all this sheathing has been removed. Everybody has become apparent. . . ." The interest and minority groups, homosexuals, women, Blacks have all demonstrated, and made demands. Further, "all of a sudden these pre-Columbian and African creations are in competition with masterpieces of the West. What could be more confusing? Who knows what is a value under these circumstances?"

Relatively few intellectuals were sympathetic to the counterculture. They were more articulate, for the most part, however, than its opponents. Most had a reputation as "far-out." One was known, however, as the personification of a conventional liberal intellectual active in politics. His position is therefore worth stating at length, since it is more sympathetic to the new culture than that held by most of his peers. Perhaps sympathy with alienated youth is easier for him because he sees developments in basically social-structural rather than in basically valuational terms — a view common to discussions about culture by the more radical. The topic he chose to talk about was the "alienation of the youth, the poor, the blacks and the low income whites."

> There are two kinds of groups, those people who have been in it and now been estranged from it, like the young and the intellectuals, and those like the Blacks, Puerto Ricans, Chicanos, etc., who have always been excluded. . . . The problem is going to require some very serious revision of structures in this country to admit where possible a much larger degree of participation. . . . A kind of moral revolution which makes today intolerable what was tolerated a generation or two ago. . . . The quickest way to stop the Black revolution would be to put the Black in every power agency in the country.

Then there were those on the Left who took a clear-cut stand in favor of the counterculture. An editor of an "underground" journal said he wanted to talk about the "repression of dissent and social change." One had to begin, he said, by "restructuring society. I mean I don't know where to take this question. . . . I don't know if I qualify as an intellectual in the sense that intellectuals agree that there should be social change, but that it should be done within the system. I think of it in terms of alternate institutions. . . . I would sooner write for an underground paper. . . . I think irreverence is the first stage of liberation."

A third view from the Left eschewed social structure but dealt directly with the quality of relations between people. A novelist explained:

> What has concerned me for the longest period is making people's lives worthwhile to them, by any imaginable device. . . .
>
> Q. What do you think man needs now?
>
> A. There are things you can give him in the next five minutes: you can give him somebody to talk to, you can give him a friendly

neighbor, you can give him somebody who's in love with him, you can give him a prize, you can give him a lot which will help right away. Maybe you have to give him a decent house. If you could do a wonderful job for one person, then maybe two and then maybe five. As President of the United States I wouldn't know what to do except to send food down to the Mississippi Delta and to clear up ringworm down there, and that sort of thing. And there are the seven spiritual works of mercy; which are to teach the ignorant, to counsel the doubtful, to console the sad, to reprieve the sinner, to forgive the offender, to bear with the oppressive and the troublesome, and to pray for all. Well, if we do that it would certainly be a great help. And the corporal works of mercy are: to feed the hungry, to give drink to the thirsty, to clothe the naked, to shelter the homeless, to visit the sick and prisoners, to ransom captives, and to bury the dead. I find this utterly beautiful and not at all utopian. So I would like to see our government, since it's in charge of this, this control of communications, to recommend these fourteen acts, including the ransom of captives, that happens to be pertinent right now.

These views of the culture crisis are not unexpectedly related to the age and circle membership of intellectuals as well as to their views of the Cold War in the 1950's and their views on the war in Vietnam. What is more, views on the culture crisis seem more related to these basic facts of intellectual life than – or certainly as much as – to views on almost any other matter taken up by our sample. For this apparently amorphous topic encapsulates more of the intellectual "scene" than almost any other issue. To be sure, what we can best predict is not whether certain intellectuals are ambivalent or liberal in their views but whether they favor the counterculture or not. Younger intellectuals tend to favor the counterculture, as might be expected, but the difference age makes is not very large – for in our group of intellectual leaders we have already seen that age is not a very important determinate of opinion. But the three members of the radical circle who chose to talk about cultural matters all favored the counterculture, while only one of the twenty-one members of the center and social science–literary circles did so. Also, somewhat predictably, only one member of these circles chose specialized topics in the area of culture, whereas five of the nineteen others chose topics off the beaten path. While half of those who opposed the Cold War in the fifties now support the counterculture, only 10 percent of the others do so. With our data it is hard to say whether attitudes toward the war in Vietnam cause one to have particular attitudes toward the counterculture or vice versa, but in any case, the two are strongly related. Supposing the counterculture orientation to be more general than attitudes toward the war, let me report that of the eight favoring students or the counterculture, six wanted to get out of Vietnam right away and the other two favored some kind of coalition. The eight ambivalents were split half and half between immediate exit and coalition, while of the twenty-one liberals, only four wanted out right away, and six favored an attempt to keep the Communists out.

At first glance intellectual prestige does not appear to be related to the stand on the counterculture, but the matter is more complicated. Proponents of different positions in the realm of culture and values found quite different kinds of intellectual atmospheres about them. Those taking either extreme – a conservative position or a radical one – tended to feel that no one else held their views and that there was in general no discussion of the entire matter. An exposé journalist who took quite a conservative position on culture and who acknowledged that few would agree with him felt, "There's almost no discussion of it. It's a question that's not too relevant." And a critic with conservative views on the culture conflict said, "I find myself going back to people with whom I had disagreed a great deal in the '50's and with whom I still disagree on many things. But they have caught much of this [his views of the culture conflict] in ways I hadn't been smart enough to see – Trilling, also Saul Bellow." These persons who had influenced did so in the past, not in the present. As for the public, as contrasted with intellectuals, he observed, "I don't know if there has been any serious public discussion. . . . Partly because the issues are still very amorphous. It will require several years of intellectual discussion of these issues before they can be presented to a larger public." A critic partly favoring the counterculture found the important influences upon his thinking even further in the past.

Most of them [people who influenced him] have been dead. . . . This is central to the issue: to what extent are we influenced by the dead? You see, the questions [about influence in the interview schedule] don't include that, because you're inclined to think in terms of issues as being one formulated by somebody in the newspapers. The whole project becomes nonsensical in so far as it doesn't take into account that all of these great issues, including Vietnam, have their roots in the whole of the modern world.

He then proceeded to name Karl Marx, Lenin, Trotsky, and Rosa Luxemburg as the great influences on his thinking.

Finally, there was the attitude expressed by one of the strong devotees of the counterculture.

There is a kind of group consciousness going, where there is no level of communication that is necessary. You know there is a kind of group objectivity, without any verbalization necessary, of what is morally right. . . . A lot of people that I talk to about it [repression in society] and am very much influenced by, I never know their names. . . . The last people that I was radically influenced by was when I was driving from Big Sur to San Francisco. I picked up two little girl hitchhikers and I can only name their first names.

In contrast, intellectuals on issues of culture and counterculture tended to feel themselves part of an ongoing discussion; their liberal position is of course the dominant one. Nonetheless the list of persons mentioned as discussion partners or as influentials on issues of culture reflected a lack of any unified concentrated center. They suggested that some of those on the right and the left who felt that there was no center to the issue were justified. On the other hand, whatever concentration there is, is evenly divided between liberal and counterculture representatives. Ninety persons were named by the over 40 persons who talked about these matters. Only 20 were named by 2 or more respondents and only 6 were named by 4 or more. Leading the list was Trilling, with 6 naming him. Paul Goodman and Marcuse each garnered 5 votes (those, as we saw, not all favorable mentions). And Bell, Howe, and Sontag – probably one of the more unlikely combinations one could make – were each named by 4 intellectuals. Goodman, Marcuse, and

Sontag have published opinions lining them up quite definitely with the counterculture. Howe's published opinions suggest he is definitely opposed; Trilling's published views suggest he does not basically like what is going on, whereas Bell, as far as I can see, has not taken a clear stance one way or the other – as a sociologist he "understands" both the counterculture and traditional intellectual culture though his own style is clearly on the side of the latter. The rest of the 20 mentioned by 2 or more are also split down the middle, with 7 favoring the counterculture and 7 opposed.[1] Once more we see that those perceived as important in a given field are persons whose views were generally more to the left than the views of those who named them.

To sum up: at the time of our interviews it seemed as if many of the intellectuals we talked with who were interested in the culture crisis were fighting a rearguard action. The "kids" had gone crazy and the intellectual Luddites were tearing down the elaborate machinery of culture which had been built up in the West since Aristotle. In one sense, however, the liberal intellectuals' worst fears seem to have been unjustified: the Radical New Left as a cultural movement has largely fizzled out. Abbie Hoffman has not captured America's campuses nor has Charles Reich's Consciousness Three become a social reality. No doubt in some small measure the resistance of major intellectual leaders to the acceptance of hedonism and anti-intellectualism helped turn the tide, but I suspect that other factors were more important. If the spirit of the 1970's is less frenetic than that of the 1960's, the "revolutionary character of our times" has not changed. Radicalism has not won out, but it remains an important force; neither are the liberals triumphant. The basic question of whither our values remains an unanswered question not to be brushed aside by technocratic solutions.

14

No Gurus Here

Aside from the pallor cast by the shadow of the war in Southeast Asia, where were leading American intellectuals in their thinking on America in 1970? What were the results of the radical sixties? Where were they going, and what were the new issues of the 1970's? Perhaps the most salient fact is that they were neither gurus nor sages, prophets nor sons of prophets. The majority were reacting to events, rather than predicting them or even creating them through the impact of their ideas. That is, the intellectual elite in their conversations about social issues for the most part had little more to say than what might be heard in the cocktail party conversation of intelligent, well-educated people. This is not to say that they could not and did not express themselves more fully in their writing in ways far more elegant than their speech, but comparisons we have made between the printed word of respondents and the statements they gave us show no substantial differences in point of view. Moreover, not every intellectual we interviewed had written directly on the topic he chose to talk about. A very influential novelist in intellectual circles told us, "I'm doing it [giving you an interview] because I don't publish views on these matters; it's not my business."

The printed word does not tell us, for the most part, the source of a person's ideas or how he happened to come upon them. For some purposes this does not matter. But it does matter that the great wave of concern about our culture was a response to events not initiated by the intellectual elite, for this allows us to predict that the concern with the culture crisis will probably fade away in the 1970's, unless some new events stir it up. The combined impact on the field of race relations of concern with the war in Vietnam and the expulsion of whites from the Civil Rights movement cannot be overrated. The intellectuals who in 1970 were still concerned with civil rights were on the fringes rather than at the center of American intellectual life. By and large litterateurs or journalists, these intellectuals despaired of any course of action or any new thought that could alter the present stalemate in race relations progress. No radical chic here, just reiterated liberalism. This impression would not have been gained, by the way, through a mere perusal of articles, books and reviews of the late sixties.

Advocates of economic and social reform at home as well as champions of peace and disarmament on the foreign affairs front had been promoting their positions for many years – generally all their adult life for the somewhat younger domestic affairs group and at least since World War II for the foreign affairs contingent. They have been leaders and influentials in the development of ideas on domestic reforms and foreign affairs. But most count their work as a failure and seem to find no way at present of carrying out their goals or even influencing others to join the battle. It would not be fair to say they are in despair, but optimism based on a step-by-step plan for achievement is simply not present.

Clearly, after the opening up of the early sixties, the lack of movement in the late sixties, together with the anti-intellectual direction of the counterculture, has left the intellectuals in disarray. It is now a time for regrouping.

Irving Howe's title for his collected essays of the middle fifties and early sixties still fits the current situation: *Steady*

Work. The issues are not especially new, even the positions are not. The problems and the necessity of a steady critique both remain. His introductory comments about radical thought seem to fit the case of almost every person we interviewed, whatever his political views.

> Now there are people who delight in pouncing upon such remarks and pointing to the difficulties, even the disarray of contemporary radical thought. They are quite within their rights. But is there any mode of serious political thought in the mid-twentieth century which is not in a condition of similar difficulty? And then there are a few old and true believers who feel it scandalous to make such admissions: as if silence or a fine show of rhetoric would veil the troublesome reality. No; candor is the beginning and the end of intellectual recovery. . . . We radicals are men like other men, bewildered, troubled, eager. . . . (Pp. xiii–ix)

IV

The Place of Elite Intellectuals in the Resolution of American Social Problems

15

Intellectuals and Power

Since Plato had his say about "philosopher-kings," the relationship between intellectuals and men of power has been of great concern to the intellectuals themselves, if not to the "kings," and has been the subject of much sociological speculation.[1] The relationship is full of conflict which cannot be solved. By definition, intellectuals consider themselves men of knowledge and wisdom, qualities best developed through contemplation and independence. Yet the intellectuals of interest to us in this book are not solely concerned with knowledge and wisdom for their own sake. Rather, their intelligence is directed toward the analysis, if not the solution, of problems of values, politics, esthetics and the human condition generally. Now, it takes a very unusual person indeed to make such analyses without the hope that he may in some way affect actual practice in the "real" world, and the most direct way of affecting the real world is to partake in some way in actual decision-making processes. Yet this very participation tends to modify the life-style that makes contemplation and independence possible.

Herman Kane and Stella Manne contributed several drafts to this chapter and have made valuable suggestions. A shorter version was read at the 1973 Annual Meeting of the American Sociological Association.

In addition to this social contradiction, there is the cultural fact that most intellectuals in modern societies tend to be "alienated" from the established powers — which means in part that the ideas and plans intellectuals tend to advocate, often leftist, egalitarian notions, are generally poorly received by men of power who respond to a more conservative public. How to solve this dilemma and what relationship to establish with men of power is a matter intellectuals have resolved quite differently in different societies. To the extent that modern societies allow intellectuals great freedom of choice, to that extent their solutions reflect their individual personality and predisposition. Thus, one finds with each society a widely differing set of approaches to the problem.

In general, there are two kinds of relations between intellectuals and men of power — direct and indirect. Direct relations involve some kind of personal contact together with an attempt to communicate an idea, point of view, or policy to a specific powerful person. Indirect relations involve writing, speaking, demonstrating, and so on. In these cases a communication is addressed to other intellectuals, or to a mass or semimass audience. If this indirect communication or attempt to influence policy has any effect at all on powerful persons it is mainly through a "trickle down" (or up, as the case may be) effect via the mass media or the even more nebulous channel of "climate of opinion." In this chapter we shall show that most elite American intellectuals, both in theory and in practice, prefer the indirect approach.

The most important structural fact about American elite intellectuals is that in the five years or so preceding our interviews (a time period covering both the Nixon and Johnson administrations) only about 30 percent had any recurring direct relation with men of power. Another 25 percent had direct relations on a very sporadic, often one-time basis. Five percent had a channel to men of power, but chose not to use it, while 40 percent had no means at all of reaching them. We will discuss the nature of these relations in the course of the chapter.

Because they did not run down to Washington with their great plans in their briefcases does not mean that the intellectuals we studied did not make other attempts to affect policy. In fact, they did attempt to make or affect opinion and thus indirectly affect policy. Ninety-five percent wrote something on the war in Vietnam and/or the issue with which they were most concerned. Almost 80 percent made public speeches on the war or the issue they discussed. Since our sample was selected on the basis of writing, the remaining 5 percent were either novelists, poets, or editors whose published literary output they themselves did not count as directly concerned with the topics of the interview. Seventy percent gave money to one cause or another under discussion. Sixty-eight percent signed petitions. Almost 60 percent joined organizations connected with the issues or with Vietnam. About 40 percent worked for a candidate or a proposal and another 40 percent participated in demonstrations. (Think of members of the power establishment taking to the picket line!) A bit less than 20 percent formed an organization. Finally, over 10 percent actually engaged in civil disobedience.

In this chapter, then, we shall note the theories on relations with power held by the American intellectual elite, shall attempt to account for why only a minority actually does relate directly to men of power, and will show what kinds of intellectuals engage in what kinds of opinion-molding attempts. Finally, we shall attempt to assess their effectiveness in these varied activities and offer some prognosis for the future.

The Role of Intellectuals in Relation to Men of Power: Their Theoretical Views

What does the phrase "men in power" mean to most intellectuals? The phrase is ours, and in questioning intellectuals we purposely left it vague so they could interpret it in their own fashion.

We obtain an even clearer notion of what intellectuals had in mind when we also examine their responses to questions relating to who the "important people" were in making policy about Vietnam or the issue with which intellectuals were themselves most concerned. It seems that to intellectuals, men in political power meant members of the White House staff, sometimes the

President (five intellectuals referred to personal conversations on Vietnam with either Kennedy or Johnson, for example), frequently other high-level officials in cabinet departments and independent agencies. But members of both houses of Congress and their aides were also included. Infrequently, intellectuals meant by "men of power" anyone who might pull some weight on an issue of interest to an intellectual. If a respondent was concerned with foreign policy, then his targets tended to be high-level persons in the administration or in Congress. A respondent interested in the state of art in America, however, would naturally have a very different list of powerful persons. While intellectuals mentioned business leaders and the military in the context of "important people" in particular issues and especially in connection with Vietnam, almost none were referred to as men of political power with whom intellectuals might have relationships. In their views of men of power, intellectuals were curiously not much different from other American leaders who also tended to focus on Washington officials.

Given that we have some idea of what intellectuals mean by men in power, there are two aspects to the direct relations of intellectuals with them. First, there are the general ideas intellectuals have about such relations and then there is the actual extent and nature of these relations. Since Western intellectuals tend to strive for logical consistency, there is a high degree of coincidence between the theories intellectuals have about power and their actual practices. Without ultimately claiming the priority of either theory or reality, let us begin with the theory if for no other reason than the fact that intellectuals are so often claimed to be men of ideas.

Because they have been said to play quite different roles in Western democracies and in Eastern societies such as the Soviet Union, it is especially useful to contrast the views of intellectuals in the United States with those of intellectuals in a socialist country with a one-party system. For the latter type of situation, we have Yugoslavia as an example. More liberal than other countries usually thought of as "East European,"[2] Yugoslavia nonetheless has had, since World War II, an indigenous Communist Party (now called the League of Communists) not imposed upon the country by outside force. Hence in some ways, intellectuals in Yugoslavia may be said to be prototypical of intellectuals in a more "liberal" socialist society.[3]

Elite intellectuals in both the United States and Yugoslavia[4] were, toward the end of their interviews, asked: "Of the following factors, choose three in order of importance to you in your work as an intellectual. Rank them 1, 2, and 3."

As might be expected, there was much complaint about the list. Many intellectuals added qualifications, comments and the like which resulted in some additional categories. Still, we obtained answers from 87 intellectuals in the United States (nonrespondents were simply those for whom time had run out in the interview or who were interviewed over the telephone) and from 98 Yugoslav intellectuals. Table 31 below lists the percentage of intellectuals in the United States and Yugoslavia who gave various aspects of their roles as the first or second most important factor in their work.

Early in our design of the study we sensed from preliminary talks with American intellectuals that they made a sharp distinction between sheer criticism and criticism which also offered solutions. We therefore added this latter alternative to those already used in Yugoslavia. The net result was that a total of 60 percent of the American intellectuals thought their role was to be critical of their society but that only half of these felt that criticism also implied making policy suggestions. While only 40 percent of the Yugoslav intellectual elite opted for criticism, the fact that they had one less choice to make in this area affected the size of their response, so that this smaller percentage should not be taken as necessarily indicating a less critical attitude on their part.

If one looks at the context of all the responses in both countries, American intellectuals are obviously less likely to give answers that reflect social responsibility. More than half want to satisfy only themselves or their colleagues. They are less interested than the Yugoslavs in formulating new goals and ideals and in reaching out to nonintellectuals and to those less aware of high culture. According to these answers, at least, American intellectuals remained more in the individualistic liberal Western tradition, while the Yugoslavs were more likely to give answers appropriate to a socialist tradition.

TABLE 31

First and Second Most Important Factors in Work as Intellectuals: United States and Yugoslav Elite Intellectuals (In Percentages)

FACTOR[a]	UNITED STATES (N = 87)	YUGOSLAVIA (N = 98)
To create works which answer to the critical standards of the intellectuals and colleagues whom you value most	46	31
To approach critically the problems of our society	31	44
To approach critically and provide solutions for the problems of our society	29	...
To communicate to people the values and ideals of our society	23	22
To formulate new ideals and goals for our society	20	39
To assure that works of quality receive the greatest publicity	9	11
To satisfy oneself[b]	9	...
To develop artistic, intellectual and cultural interests among the people	8	19
To create works which the majority of people can understand and like	7	32

[a] Listed in order of importance in the United States – not actual presentation on checklist.
[b] Added by 9 percent of the United States sample.

Some American intellectuals were more interested than others in satisfying their peers. As might be expected, 80 percent of intellectuals who are professors in the humanities chose satisfying other intellectuals; more than 40 percent of the social science professors said their role was not only to be critical but also to solve problems. Interestingly, being critical (but not necessarily being critical and trying to solve problems) was a role chosen more often by intellectuals of higher prestige.

Further, most American intellectuals do not strongly believe they should even have close relations with men of power. Only 30 percent of those we talked with on this subject thought intellectuals should have some relationship with men of power, another 30 percent or so thought they ought to have a relationship but only under carefully specified conditions; a few were ambivalent and about 25 percent firmly said no.[5]

Aside from the "oughts" of the matter, about half thought intellectuals actually did have relationships with men of power in 1970, while about half thought they did not. At the very most, therefore, intellectuals were ambivalent about relating to men of power. At least half of the affirmative remarks were so heavily qualified so as to leave the nature of the relationship very much in doubt. The qualifications and the hedging are more important than the statistics of response, so let us look into the answers in more detail. A minority thought involvement with power and policymaking was obligatory. For example, a journalist said, "They [intellectuals] have to be involved in policy. Everybody's got to be involved in policy that has any stature." He then attacked the "ivory tower" image of intellectuals. "When I was much younger," he said, "the professors would say that it wasn't their function to see that ideas were disseminated beyond a certain point." Others thought involvement was essential to an intellectual's most precious quality: understanding. "One preserves one's security in distance from men in office at the expense of one's understanding."

Still others endorsed some form of participation with men of power, but did not say it was obligatory; they left it, instead, as a matter of individual preference.

> I think that there are plenty of intellectuals who want to maintain a distant relationship, who want to be called to give their ideas [and yet] be free and be critical.

The critical role of intellectuals came up in the comments of many of those who qualified their assent as well as in the responses of those who flatly opposed a relation between intellectuals and men of power. In fact, two thirds of those 39 who gave any reason at all for their answers to this question mentioned the issue of the critical faculty of intellectuals, and half of these thought it was simply impossible for intellectuals both to participate in policymaking and to remain critical. For example:

> In most cases where they tried that [participation in policymaking] it ended in disaster. Their major role is to be critical. They are men who have to stand for more general standards and therefore they will always clash with the politicians, who must always think in pragmatic terms. And they are the conscience of society. In most cases they have become coopted and they have abdicated their main function, the critical function. So in most cases they have become what I call mental technicians.

Other intellectuals who spoke about the critical role of intellectuals felt that the relationship between intellectuals and men of power was inherently contradictory:

> The two are really incompatible. . . . In order to preserve his political effectiveness he [the intellectual] is inhibited from saying certain things. . . . At the same time commitment vitiates the full exercise of his intellect. Once you've gotten committed to something, you can't really think in an absolutely free way about it.

Not all intellectuals concerned about being critical felt that contact with power was necessarily destructive of their traditional role:

> You get the argument that an intellectual is always supposed to be critical. Now there are two things involved with that. One is the misuse of the word "critical." The original use of the word is not to be negative [but] . . . to explicate an underlying structure. . . . [Second,] if you are critical, at what point do you take the responsibility for what you have said?

Several intellectuals who had served the government felt that taking responsibility was paramount and that this was a way of preserving criticalness:

If things are done of which you very strongly disapprove, not necessarily on a moral ground, but even simply wrong policies, then there is no alternative to breaking with the Administration, with power, and speaking out.

The issue of being critical was, of course, related to an intellectual's general view of his role; those who favored criticism *and* policy solution were, first, more likely to *want* to relate to men of power and, second, less concerned about the possibility of losing their own critical faculties.

The critical quality of intellectuals was not the only issue in the discussion of the relations of intellectuals to men of power. The very nature of the role itself was debated.

Most intellectuals ruled out actual decision-making roles but accepted the idea of advisory, consultant or expert roles. As one might expect, the sociologists in our sample were articulate about matters of role. One, who specifically opposed the idea of "philosopher-king," did feel nevertheless that intellectuals should occupy official advisory positions, such as in the Council of Economic Advisors and the "Council of Science Advisors." But while some respondents drew fine lines between an executive position, an appointed position for doing a specific limited job, an official advisory position and an unofficial one, another sociologist explained that in practice it was often hard to draw the boundaries between advice and policymaking.

You see then, there are two things: the expert role and the advocate role. All sorts of academics and intellectuals are consulted as experts and as social "engineers." Take the Vietnam thing. Suppose he is called in to advise on Vietnam. Now if you're committed, I would guess that most academics who, say, advise on Vietnam, advise as partisans – even though many do not. Today maybe that's a bad example, because there is so much emotionalism – people are so polarized.

So that this whole question of when you're actually the scholar and when you're acting as a politician is a very curious thing.

. . . If you take the Civil Service – typically the Civil Servant is supposed to be neutral and when he's asked by his minister for a memo on some subject, he's supposed to give the facts and not where he should be going. But the practice and studies have shown that bureaucracy makes policy. And the same way, advisors make policy, at least influence it.

Actually, only five intellectuals we talked with endorsed out-and-out participation in actual decision-making activities.

All this discussion of the role of intellectuals presupposes that in fact they have something important to contribute to men of power. Surprisingly, a good number of the elite intellectuals we interviewed denied that intellectuals had anything at all to offer. One of the stronger statements among those opposed to relationships with men of power suggested that intellectuals might actually be a menace to a society.

> As a rule, I have a profound distrust of intellectuals because I think historically they have had a fascination for the authoritarian and the grotesque ideologies and mythologies that tend to subvert Western Civilization. I think the question is: Is society endangered by the intellectual if he gets too close to power?

The very large majority of the intellectuals we spoke with, in sum, were not only interested in a more individualistic role but were also quite wary of involvement with men of power.

Direct Access to Power: The Reality

All this is in theory. What is the reality? We saw that relatively few intellectuals had much contact directly with men of power. Let us see just what this contact was like.

There were four ways leading intellectuals related to men in power in political life. One was directly, either as an "insider" or as a person outside of government or politics but with some other means of direct access. Insiders either held government positions or had recently held such jobs.* For example, a former member of the Council of Economic Advisors took full advantage of his special channels.

* This classification includes jobs or channels in the Johnson administration so that we do not reflect wholly the Republican tendency to make less use of intellectuals.

I do have some channel directly to people in Washington but they tend to be specialized in the economic line, but I do push on them. . . . And whenever I have a conversation with people I know in government – like Pat Moynihan, for instance – I make no bones about my feelings.

Others who were outside government were like the social scientist who told us he had "been working with Congress a lot" and had "done seven hearings in the last year." He noted, "I can reach almost anybody in the world I want to reach with a telephone, you know. And I use the telephone." Altogether, 30 percent of the 90 intellectuals in our sample who gave us complete information about their influence activities had these direct channels either on Vietnam or the issue with which they were most concerned.[6]

Another 25 percent of our sample of elite intellectuals had irregular channels to government or political leaders either on Vietnam or their "issue." The role of key intellectual "brokers" who are themselves insiders with regular channels was obvious for the first group, but for occasional communicators with power their importance cannot be overestimated. For example:

I have very little experience with reaching directly into these circles. Nevertheless, one sometimes has the chance to speak to power brokers as a middle man. Last week I had a chance on some very special issue to meet with Pat Moynihan.

Then there were half a dozen or so who did know key intellectuals in official positions or other men of political power but for one reason or another were reluctant to make use of these contacts.

I know various people. I know Kissinger and I know various Senators and I know a couple of people in the House of Representatives and some people in government. But it doesn't really work that way.

Usually the fact that an intellectual had these contacts was well known to his friends, who may have urged him to mobilize them.

I know people who talk to the President. I can talk to people who do talk to him. A lot of people want me to take the initiative and say different things to them.

Finally, almost 40 percent of the elite intellectuals had no channels at all to men of power. Some admitted, "I wouldn't know where to begin." Others thought it was "a waste of time." A few had some indirect connections but did not know how to mobilize them. Finally, a radical position taken by someone who felt quite out of it was that he could reach important people only "by endangering their position, by threatening the things that they want to maintain by blackmail, in other words." Not quite the kind of communication we had in mind.

More conventionally, for some intellectuals, contact with men of power, "important people" meant merely communicating their views. A foreign correspondent, an expert on Vietnam, explained:

I've tried to do it on and off through the years. . . . I see them when I go down, when I come back from the country. I usually come back once a year and go down to Washington and spend a week or so and manage to see most of these people and give them my views.

For most other intellectuals, however, reaching people involved some indication that their ideas were taken seriously. For example, take this account of attempted influence with regard to the war in Vietnam.

Well, I would work. I would try to strengthen the hand in the House and Senate [of those] who were taking the kind of position that I think ought to be taken. I would feed them information. I would encourage them to go further than they have gone. I would do everything I could to support them. . . . For example, I thought one of the most significant things that had happened in reference to the Vietnam War from a domestic-political point of view was when Senator Thurston Morton, who had been a hawk all along, suddenly decided that it had gone too far. . . . Well, when Morton switched, we praised Morton and tried to do everything we could to make his views as widely known as possible.

For still others, influence meant a fusion of the communication and action functions into a general style that ultimately had some effect, direct or indirect, on the thinking of policymakers. The following comments (made in 1970, remember) illustrate the blending of the two functions.

> Well obviously Moynihan . . . does play a particular kind of role in affecting what kinds of things are thought about. Let's say his work in urban affairs, and after all, if he says let's have this kind of educational act, we're going to change educational policy; or, let's treat minority groups in a certain way, then this indeed has an impact — not again necessarily on money priorities, but on the way people in the central policy making area do tend to think about things. And that might in fact affect the disbursing of funds as well. . . .

Not surprisingly, an intellectual's experiences with men of power fairly well match his views and reflections about these experiences. Seventy percent of intellectuals with regular channels to men of power thought intellectuals *ought* to be involved; less than 15 percent of the rest thought so. As to their perceptions of the facts of the situation, almost all those with regular channels felt intellectuals had some influence, though many credited the Kennedy administration for an increase in the influence of intellectuals. Others were less positive about the degree to which intellectuals were influential, and even those with regular channels thought Republican administrations in general and the Nixon one in particular to be impervious to intellectual influence. In their view, there are no ". . . effective means for communicating with the President or his advisers. I mean, you're talking about me personally doing something. I mean, I really can't conceive of anyone having an influence on the President of the United States."

And there was some degree of sour grapes among intellectuals who had irregular contacts with Washington officials. One recounted this story:

> I went to Washington to talk to Kissinger about Biafra. I actually flew to Biafra and spent a week there. And then . . . I went to Washington to lobby personally and saw two people. I saw Kissinger and I saw . . . Mike Mansfield. . . . My chief feeling was impotence. It wasn't that they were hostile or unfriendly or even in the massive case unconcerned. It was that . . . in order to do anything,

you'd have to make a monumental stand or effort which we weren't ready to do. And I certainly walked out of there feeling baffled and defeated – the sense of not knowing where to really put the pressure on.

In sum, only a minority of elite intellectuals have regular access to men of power, though over half manage at least irregular contact. When they do get to see them, they often try to do more than merely state their views, though most either experience frustration or even without having tried at all feel the attempt would have been worthless. Much of the contact seems to take place through the official White House intellectuals. All of this certainly does not represent total isolation from power; neither, however, is there an image of comfortable and frequent intimacy.

All this is from the point of view of the intellectuals themselves. The reluctance of the kind of intellectuals we had been interviewing to become engaged with men of power is matched only by the low possibility that men of power themselves are indeed influenced directly by such intellectuals. In the Columbia University study of American leadership,[7] we asked in 1971 for each respondent to give the names of persons with whom he had talked, who had good ideas, who in other ways influenced him or whom they thought influential on such matters as were being discussed (for a list of the issues in order of importance, see Page 241). Very few indeed of those who intellectuals themselves thought were important or who in any way were similarly named by intellectuals were also named by the rest of the American leadership. The list below includes all persons mentioned five or more times by the American leadership who were either interviewed as intellectuals in our study – only eight, though another six were on our list to be interviewed but either refused us or could not be scheduled in time; or who wrote at least two articles or whose book was reviewed at least once in the list of leading intellectual journals compiled for the purpose of drawing our sample of intellectuals (see Appendix), or who were mentioned in any connection by at least two intellectuals during the course of our interviews with them. Persons who obviously had only political roles such as J. F.

Kennedy, Fulbright, Nixon, and so on were omitted. Generously, one might call the persons on the list "intellectuals," for they meet the standards we used to draw our sample, but at a lower level – for example, two articles rather than four, and so on. More precisely, they are persons who are at least *visible* to intellectuals on matters of social policy.

This is obviously a very different sort of list from that of the top 70 intellectuals given in Chapter 1, or for that matter from our sample. Of the 55 persons on the list here, we tried to interview 14 as elite American intellectuals and succeeded in reaching 8. Names marked with an asterisk are listed in the top 70 as well as in this tabulation, and there are only 7. The crosses show persons named by intellectuals as Vietnam experts and the circles show those named additionally by at least 5 intellectuals as experts on some issue or other of concern to the respondent, other than Vietnam. There are only 19 marks in front of the 55 names below. In the other direction, in addition to the 7 already noted who were on the list of the top 70 intellectuals, there were others who were named by men of power but not often enough to make this list. (Each member of the bottom rank was named 4 times.) David Riesman and Arthur Schlesinger were named three times by the men of power, Buckminster Fuller, Norman Mailer and Hans Morgenthau were named twice, and 7 others were named once. The other 51 of the top 70 intellectuals were not named by men of power even once. All this does not suggest a very high degree of direct influence by the elite American intellectuals on men of power in America.

A close inspection of the kinds of person named by intellectuals as compared with those named by men of power is even more revealing. The men of power tend to name technicians, intellectuals who have held office, some reporters and columnists, mass media "personalities," and others whom Lewis Coser called intellectual "celebrities."[8]

We must conclude on the basis of the reports of both intellectuals and men of power that there is very little direct cross talk between the two worlds.

The type of person named by men of power begins to give us some clue as to why intellectuals do not directly relate to men

The Fifty-five Influentials

RANKS 1–10

Burns, A. F.
° Friedman, Milton
* Galbraith, John Kenneth
* Gardner, John
° Kissinger, Henry
* Moynihan, Patrick
Nader, Ralph
* Reston, James
Wilkins, Roy
Young, Whitney

RANKS 11–20 (TIES)

Acheson, Dean
† Alsop, Joseph
† Brinkley, David
° Buckley, William
† Bundy, McGeorge
Commoner, Barry
Ellsberg, Daniel
† Halberstam, David
Heller, Walter
† Huntley, Chet
° King, Martin Luther
McCloy, John
McNamara, Robert
Samuelson, Paul
Sevareid, Eric
Smith, Howard K.
† Wicker, Tom

RANKS 21–24 (TIES)

Brewster, Kingman
Bundy, William P.
° Clark, Kenneth
Clifford, Clark
Dale, Edwin L., Jr.
Evans, Rowland
Fairbank, John K.
† Fall, Bernard
Freeman, Orville
Goldberg, Arthur
Hoffer, Eric
* Kennan, George
Kerr, Clark
* Lippmann, Walter
Lowenstein, Allard
Novak, Robert
Okun, Arthur
Pechman, Joseph
Raskin, A. H.
Reischauer, Edwin
Rostow, Walt
Rusk, Dean
* Rustin, Bayard
Shakespeare, Frank
Sheehan, Neil
Sulzberger, Cyrus L.
Tobin, James
White, Theodore

NOTE: Influentials are: (a) Those interviewed as intellectuals; (b) Those who wrote two articles or whose book was reviewed once in our master sample of leading magazines, 1964 through 1968; (c) Those who were mentioned in any connection by two or more intellectuals interviewed in 1970.

* Was on list of top 70 prestige intellectuals
† Was on list of top Vietnam influential intellectuals
° Was among top intellectuals mentioned on issues (five "votes" or more)

of power. Analysis of the characteristics of intellectuals who by their own account have channels to men of power, as compared to those who do not, further suggests that the general lack of relationship is a matter of the style of intellectuals, the fact that most intellectuals in our sample are not technical experts and the fact that most do not serve government in the first place.

Style is perhaps most important, for one's general intellectual style underlies the theories, the perceptions and the actual involvement of intellectuals. The matter is best explained in Podhoretz's account of how the "family" of New York literary intellectuals reacted to the Kennedy administration.

> The family did not regard certain of these "leading representatives" who were being honored with invitations to the White House and appointments to government task forces as intellectuals at all – they were, the family said, academics or high-level technicians – and within the Kennedy administration itself, only Schlesinger, Galbraith, Goodwin, and Moynihan really were acknowledged as intellectuals in family eyes. . . .
>
> Sometimes we were consulted directly – and made extremely uneasy by the kind of thing we discovered was expected of us. Once, for example, I myself was summoned to Washington by a *very* [one supposes he means Kennedy himself] high member of the Kennedy administration who wanted to know my "ideas" (that was the word he used) about the situation in Harlem.
>
> I had, as it happened, a great many of what I would have called "ideas" – and interesting ones as I thought – on this subject, but I noticed while expounding them over a very good lunch that the great man was growing restless and bored. This puzzled me, for I thought I was speaking well. Faltering a bit in response to his impatience, I asked him whether he disagreed with what I was saying. "No, no," he answered, "what you're saying is all very well, but what should we *do* about it?
>
> *Do?* I was not accustomed to thinking in such terms; I was accustomed to making critical analyses whose point was to understand a problem as fully as possible, not to affect or manipulate it.[9]

If one is a general critic one simply has not much to say of interest to men of action. Policymakers need specific advice on technical matters, not general theory.

Intellectual style includes not only a propensity for general theory but an intellectual's values, opinions, specific interests, as well as his general views of his role. Thus, there is a relationship between an intellectual's perception of his general role in society and the likelihood that he will have direct access to "important people." The majority of those who felt that an important role for intellectuals was to be critical *and* to offer solutions had regular channels to men of power. Only one third who stressed the critical function *without* seeing the need to offer solutions had access to power, and only one fifth of those who felt intellectuals should be concerned with satisfying other intellectuals had special channels.

Ideas were important in still another way in affecting who became involved in government and national power. Those in our sample with strong left-wing or radical ideas simply fell outside the range of permissible differences of opinion. The relatively small number of radical left-wing intellectuals in our sample were not likely to have access to power circles even if they fulfilled other criteria – that is, even if they had technically useful occupations and were concerned with the "right" issues such as foreign policy or domestic reform. These intellectuals clearly would have liked to influence events but their solutions were not in demand. One suggested that men of power could not be reached "unless you kidnapped them, that would be the direct approach!"

Then there are those who may not have been permanently barred but who happen to have held an incorrect "line" at a given moment. Many of these were persons whose "line" was correct in some previous administration but whose ideas were now out of style. Because almost all intellectuals vote Democrat (see Chapter 1), this happens more often with Republican administrations, as respondents themselves noted. Then there was Vietnam, an issue which transcended party line in that most intellectuals who might otherwise have had channels to power felt quite impotent on this matter. The following was typical:

In 1965 after Johnson began the escalation – at that time I had friends in the Democratic administration in Washington – I talked to one of them. I said, "Jesus, can't anybody turn this man around? Can't anybody talk to him?" He said (the friend in the administration), "I'm sorry, there's not a chance."

All this suggests that not only what an intellectual thinks, but what he thinks *about* affects his relation to men of power. Concern with "hot" issues such as Vietnam do bring an intellectual into the field of vision of the powerful. So despite the fact that some of the more prominent Vietnam experts were barred from Washington power circles because of their undesirable views, almost half of those nominated two or more times by other intellectuals as important on Vietnam or as discussion partners on the subject had direct regular access to men of power. Less than one quarter of the others has such channels. Of course sheer visibility was important in getting a high Vietnam "vote," and visibility is both the result of and the cause of political involvement.

The likelihood that an issue of concern to an intellectual will lead him to regular contact with men of power is increased by the political potential of the issue and, perhaps more important, by its technical, "expert" character. Thus, 80 percent of intellectuals who indicated, even after talking about Vietnam, that foreign policy was their special interest had regular direct channels to men of power. And these were of two distinct types: experts in foreign affairs who were regularly consulted on matters lying within their professional specialties and who were either academics or past government officeholders or both; and special issue lobbyists. Most of the lobbyists were traditional liberals who had responded to the threat of the Cold War by actively campaigning for disarmament. They were founders of such organizations as SANE and believed that the way effectively to influence policy was to be strong, well organized and regularly in contact with inner circles of government. This was a very important group whose effectiveness somehow was decreased by the war in Vietnam.

Then there were ecologists and experts on a variety of miscellaneous topics such as foreign travel, aid to the arts, and so on, who specialized in the area of interest to them and who again were quite likely (70 percent) to be involved in regular relations with government notables.

The domestic reformers were a mixture of experts (for example, economists) and generalists whose concerns were with the social, political and economic system as a whole. The former were likely channel users, the latter were not. This division helps to explain a split within the reformers between those who used special channels (one third regularly, one quarter occasionally), and those 40 percent who had no access.

On the other hand, intellectuals whose subject was youth and values dealt with an extremely general issue, an issue that has long been a primary concern to intellectuals – the state of moral and cultural values within the United States. Those specifically concerned with campus revolts tended to cast their views in general value terms too. In any case, the concerns were not really policy-oriented and only 20 percent had regular access to channels, and these because of some other interest or their sheer prominence. Those who discussed values generally indicated a broad sense of malaise, but little that could be construed as intended for policy implementation. The problems that they saw could not be alleviated by the passage of an act or the appropriation of money. For the unhappy professors, the concern was essentially a matter for the universities themselves. It was a case of academic institutions putting their own house in order.

Finally, the group that had least access to channels was that of the race activists. Only 10 percent were regular channel users; 60 percent had no access at all. As with the respondents who discussed values, the intellectuals concerned with race were also generalists. The lack of channels reflected the dearth of new ideas put forward to help solve the ongoing racial problems of the country, and the very painful failure of those who at one time had been in the forefront of the racial crusade.

Perhaps one of the more reliable indicators of a certain style of technical competence together with an interest in active politics is occupation. Over 40 percent of professors not in the humanities had regular channels to power, but only a bit over 10 percent of the professors in the humanities had such access.

The professors of literature, English, history, art, and the like are, of course the classical "literary intellectuals," generalists in matters of ideas and certainly not technicians. The political scientists, sociologists and economists in the social sciences had a potential technical expertise to communicate and some were inclined to communicate not only to their colleagues but to men of power as well. This inclination may have been helped along through recruitment by other intellectuals. A typical story was told by a theoretical economist:

> I spent [years] working at the Council of Economic Advisors. I did that because the people who came with Kennedy into the economic advice business in Washington . . . asked me to join them. I had been in my own mind quite clearly opposed to much of the economic policy line of the Eisenhower administration and finally decided I ought to put my money where my mouth was. . . . I've come away with a permanent interest in doing the kind of research and work that is much more directly linked to economic policy systems . . . than it used to be.

One of the reasons social scientists and other professors not in the humanities became involved in high-level channels is that, unlike this economist, they were already engaged in various consulting, policy advice, and political activities. Thus two thirds of these professors were, in addition to whatever active channels of influence at top level of government they may have had, already involved in at least one current policymaking activity. Only 30 percent of humanities professors were so involved. Sixty percent of nonhumanities professors in our sample were involved in two or more consulting activities, whereas only 40 percent of the humanities professors were so involved. More than half of those with at least one policy advice activity had direct regular channels as compared with almost none of those who do not engage in policy advice; half of those with two or more consulting activities have regular channels as compared with 10 percent of those with one or no consulting work; and 40 percent of those engaged in at least one political activity have regular channels as compared with 10 percent of those who do not.

Thus, there is a large degree of overlap between the consultants and the social science experts. Between them, they make up the largest proportion of regular channel users, though they are a minority of all intellectuals.

A past involvement in policymaking also leads to current involvement. Such an intellectual has a history of prior political work. He has been a consultant or adviser to government, has worked in a cabinet department or other executive agency, or has been a member of the White House staff. He may have been involved in these activities as an expert of the type discussed above, or he may have been a one-time civil servant; a generalist who nonetheless, as part of his work, had regular access to important political leaders and has retained that access though he is no longer in government service. He may also have been a skilled administrator or political partisan called upon for service to a particular presidential administration. Thus, more than 60 percent of the twenty-one intellectuals in these categories still had regular access to channels. Once involved in the "policy mills," most of these men have retained entry, although the degree of such entry may vary from one administration to the next.

Aside from the professors, two other occupational groups bear an important relationship to men of power: the editors of prominent journals, and the well-known newspaper and periodical reporters. The editors are a mixed bag of prominent and not so prominent intellectuals. Their degree of participation in policy circles also appears to be varied, with several standing out who might be described as activists pursuing contacts with men in powerful positions. In this instance, the high level of activity appears to be less related to their roles as editors and more to their specific interests in foreign policy generally, and the Vietnam War specifically. We shall discuss this concern with foreign policy in greater detail below, but it originated as a negative response to the Cold War. It manifested itself in very serious concerns in the area of disarmament, with an agreed-upon outlook that the way to move things was through the system, and so organizationally and individually, these editors and journalists have tried to make their mark.

The reporters generally, along with some of the editors, value their "objectivity" and their roles as fact finders and information dispensers – the latter function to be performed essentially for the public. Objectivity here is defined as not taking a position.

As such, they are loath to view themselves as having influenced their sources. Even a radical correspondent felt this way:

> If you're a reporter and you take your business seriously you can't do that kind of stuff [participate in policymaking – he had in mind a specific committee whose meetings he was invited to attend]. . . . You compromise your independence. So I'm pretty careful. That's regarded as being a "pig" in the radical community because it means you're not an activist. . . . I don't have to try because I'm very interested in part of these things, but I try to stick pretty independent so I don't participate.

This all comes back to intellectual style. As we have seen, this is hardly entirely a matter of personal preference, but something built into the norms – indeed the very basic values of different kinds of intellectual pursuits. And most of the norms seem to run against active participation in policymaking, or at least they do for the time being.

Indirect Approaches to Power

Though relatively few elite intellectuals had direct access to men of power, this does not mean that the elite intellectuals were totally politically inert. Most of their activity was indirect. Here we will discuss their political and organizational activities on behalf of their position on Vietnam and on the issue with which they were most concerned. In the final section we shall attempt to assess the consequences of their most characteristic modes of political behavior: writing and talking.

The extent to which an intellectual was politically active was determined: by whether or not he had a direct line to men of power (if he did, he was less likely to engage in demonstrations and never engaged in civil disobedience); by his ideology (the more Left, the more likely he was to do the very things the establishment intellectuals would not do); by his ethnic-religious tradition of political activism (Jews were frequently more politically active); by his occupation (a social scientist may have become technically involved, an activity which then led to his political involvement); and finally, because all these factors interact, by his place in a network of other interested persons who because of their interests tended to dragoon him into political activities. Perhaps most important, the more prominent and visible an intellectual was, the more he either initiated political activity or was dragged into it. In this sense, the more a person was a prototypical intellectual, the more likely he was to be politically active. Table 32 contains a list of political activities engaged in by the intellectuals in our sample, arranged according to the difference it makes if one has regular channels to men of power or if one does not.

Clearly, demonstrations and civil disobedience are the methods of "influence" of those without other means of relating to men of power. The differences on most activities are even greater when those with no channels at all are compared to those with regular ones. Nearly 30 percent of those with no

TABLE 32

Intellectuals' Engagement in Political Activities, According to Their Access to Men of Power (In Percentages)

	INTELLECTUALS' ACCESS TO MEN OF POWER	
ACTIVITY	REGULAR (N = 27)	NONE (N = 62)
Worked for political candidate or proposal	48	40
Joined an organization	63	61
Formed an organization	15	21
Gave money	70	76
Signed petitions	63	69
Participated in demonstrations	22	50
Engaged in civil disobedience	0	19

channels (35) engaged in civil disobedience; almost 80 percent formed organizations, and so on. On the other hand, it is also true that intellectuals with regular access to channels do not drop other political activities. For much of the organizational work there either is no difference between those with channels and those without, or else those with channels prove to be even more active. Lobbying is an organized activity in the United States, even for the intellectual elite.

But apart from access to channels – which is in itself related to political ideology, with the liberals having much better access than the radicals – ideology makes a further difference. Except for the establishment practice of supporting candidates or working for official proposals, those whom we judged to be radical were far more likely to engage in political activity than any others among the intellectuals.[10] And this tendency is of course especially marked when it comes to demonstrations and civil disobedience. Eighty-five percent of the 14 radicals demonstrated, as compared with 35 percent of the 52 liberals. Over 40 percent of the radicals engaged in civil disobedience, as compared with 6 percent of the liberals.

The occupation, religion, and prestige of intellectuals also importantly affect their political activities, and these characteristics are all intertwined. To begin with, social science professors simply behaved differently on the war in Vietnam from other intellectuals; they were twice as likely to join organizations and much less likely to engage in civil disobedience. And professors in general, as compared with nonacademics, were more likely to give money and to sign petitions. Roughly the same holds true for the issue of greatest concern to intellectuals, but the trends are less strong.

All these political activities on Vietnam were also more characteristic of Jews, and there were more Jews among the professors than among the nonacademics. That is, Jews were more likely to join organizations, give money, sign petitions, and engage in demonstrations. They were even more likely to support candidates or propositions. When we put it all together – that is, look at both religion and occupation simultaneously – it turns out that the Jewish social science professors were especially different from the rest of the intellectuals. Almost 85 percent of the Jewish social science professors joined organizations, as compared with a bit more than 25 percent of the non-Jewish social science professors. But the 25 percent joining rate is the same for Jewish and for non-Jewish professors in the humanities. Apparently, organizational life simply does not appeal to humanists. Jewish nonacademics are more active than the other nonacademics, with over half joining organizations, as compared with about 25 percent of non-Jews. But it is mainly the Jewish social scientists who are especially active. More or less the same is true for giving money: 100 percent of the Jewish social scientists gave money for various Vietnam War causes, but there are no differences between Jews and non-Jews for members of other occupations. Whatever their religion, the nonacademics were least likely to give money.

It seems that for the more passive organizational activities connected with Vietnam, such as merely joining an organization or giving money, there was a network of mainly Jewish social scientists which to some extent spilled over onto nonacademic Jews but not onto non-Jews. The network theory is made plausible by the jumbled set of findings we get on specific issues other than Vietnam, where for each issue and for each set of combinations of occupation and religion we get different results. Recruitment to organizations connected with these issues occurred over a longer period of time and happened in many different ways, so that whatever networks there may have been probably worked at cross purposes.

On more active matters such as demonstrations, petitions and civil disobedience, inner disposition seems more important than a network of influence. Jews, whatever their occupation, were more likely to demonstrate and to sign petitions. In the latter activity, it is true that social science professors dominate, but it is mainly because 11 of the 12 Jewish social science professors had signed petitions, as compared with only 5 of the 10 non-Jewish social science professors.

Then there is the matter of really sticking one's neck out — civil disobedience. Religion is not especially important here — this is really a matter upon which norms had not been related to traditional patterns, for there has not been much of a tradition of civil disobedience among American intellectuals. But one striking fact stands out from all the rest: not one of the social science professors – Jewish or otherwise – engaged in civil disobedience. Their sense of civil orderliness, which may be attributed to functionalism or Keynesianism or some other theory of social order, must in some way be responsible for this finding. And it is this lack of lawbreaking which must lie behind the charge of many radicals that leading American social scientists are essentially conservative. The truth is that overall, in comparison to persons in other disciplines, social scientists are much more to the left in their opinions. But opinion and organizational activity are apparently quite different from deliberately breaking the law in order to make a point.

From all that has been presented here, it seems that elite intellectuals are close to being completely ineffective: they failed to take a moral or ideological stand on the war, and it continued; they had little new or good to say about the issues with which they were most concerned; and then finally, only a minority had any effective channels to men of power. The data we shall now discuss redeems this picture to some degree.

The more prestigeful an intellectual, the more likely he was to engage in political activity. Less than half of those intellectuals named once or not at all by their colleagues in connection with any matter except the war in Vietnam had joined an organization on the war issue or in connection with the issue with which they were most concerned; almost 60 percent of those named two to five times and 70 percent of those named six or more times had joined organizations. Similarly, 66 percent of those infrequently named gave money to some political cause, but 80 percent of those most often named did so. About 30 percent of the infrequently named demonstrated; 45 percent of the most often named did so. Petition signing shows a striking relationship to prestige with half the infrequently named but 80 percent of those most often named signing petitions. Even civil disobedience is related to prestige, despite the impression of some intellectuals that the big names ducked out: 10 percent of those infrequently named reported such activity as against 20 percent of the most often named. With some minor discrepancies all these findings hold true when occupation is taken into account, even though professors tend to have higher prestige than non-academics.

How much did sheer visibility contribute to this phenomenon? There are three possibilities: high-prestige intellectuals participated more because political activity is the essence of being an elite intellectual and the more elite the more participation; or high-prestige intellectuals participated more because their names were better known and therefore activists were more likely to try to recruit them to their schemes; or political activity made an intellectual more well-known and therefore more likely to be named by other intellectuals in one connection or another. This last explanation seems the least plausible, since with the exception of civil disobedience, and perhaps petition signing (if published) or getting one's name on organization stationery none of the activities is especially likely to lead to increased visibility. It is writing which makes an intellectual known among other intellectuals, and writing in turn is related to all of these activities. It is hard with our data to choose between the first two alternatives. Most important, however, whether intellectuals of high prestige engaged in political activities because they were asked, or whether they initiated such action because they were predisposed to do so, the net result is that a norm is affirmed: elite intellectuals *ought* to participate in political activities (other than direct policy advice). This norm of participation is even more important since sheer intellectual prestige outside of the Vietnam issue was *not* related to having good access to channels. Now it is evident why we think the intellectual elite as a group are politically active – the more "intellectual" a person is in the eyes of other intellectuals, the more he participates in political activities.

Man of Words

The main characteristic of the intellectual's role is not advising men of power nor participating in political activities but using words. Writing and speaking were one of the main activities of intellectuals on behalf of their positions on the war in Vietnam and on the various issues of special interest to them.

This activity may have had an indirect influence on policies through a trickle-down effect and as we saw in Chapter 2 when we discussed journals, this effect was believed in by a good number of editors of the leading intellectual journals. A leading writer, once a Communist in the 1930's and subsequently a celebrated anti-Communist – we say this to suggest that he has been around a long time – developed the notion of a climate of opinion.

> I think the intellectuals have an indirect influence very often by creating an atmosphere that may permeate the centers of power. Every once in a while, I suppose, there is an example of a direct influence . . . [but] it's very hard to trace that influence . . . they have at best only indirect influence which you can only guess at and never really be sure.

An editor with long experience in attempts to influence key persons in Washington and whom we classified as having regular access to channels, developed despite this access a more general theory of influence, again noting how difficult it was to account for change.

> I think we take too narrow a view of politics in this country. Most people tend to think of politics as something that happens every four years when you vote Brown vs. Jones. To me politics is a much more comprehensive concept than that. It involves conversation, the expression of attitudes, activities, all kinds of things. Now the interesting thing to me is that after an issue has had a certain amount of time, people become aware of it and it begins to saturate their consciousness. Then suddenly there takes place a change of the structure of attitudes. Whereas one day in the past support was shown for a policy, suddenly the support is not there. Now, how did that change come about? Well, I think it is very difficult to pinpoint any one thing.

How accurate are these perceptions that somehow somewhere intellectuals, through a trickle-down effect have some say on the course of events in this country? We too cannot pinpoint the effect of any single effort, at least not within the framework of the present study. But we can demonstrate that a considerable proportion of the most important persons in the mass media do read at least some of the leading intellectual journals and that a significant number of leaders in other sectors of the elite, notably Congress, also read these journals. (The feeling of some of our intellectuals that they reached Congress does appear to be correct.) Whether they have any effect is another matter, but they are read, as Table 33 shows.

To be sure, the order of size of audience is somewhat different in this list from the order given in Chapter 2. The *New York Review of Books* is quite far down the line, for example, though even here, 60 percent of the omnivorous mass media leaders are likely to read it at least occasionally. Twenty-five percent of the politically appointed federal government top administrators and White House staff read it, for these appointees appear to have the most "intellectual" tastes of any of the American leadership. In any case, this may seem to be "a lot" of readers, considering what we might expect. It is also worth noting those magazines read almost equally by the mass media leaders and by at least one other sector of the elite, for these are the journals which have the most direct effects, if any can be said to exist.[11] The *New York Times Magazine,* read by almost all the mass media leaders, also seems especially effective in reaching top government officials, though it does less well in Congress. The *New Yorker* is in a class by itself, read by almost all the mass media elite *and* by almost all of those few persons of overwhelming wealth that we could reach. The *New Republic,* widely read by the mass media leadership, also scores relatively well with labor, as does the *Nation,* which however is not widely read by the mass media elite. Congress's favorite intellectual journal is *Saturday Review,* which does almost as well with them as it does with the mass media leaders. The *National Review* does relatively poorly with the mass media men, but quite well with Congress.

The general impression, then, is that the more "intellectual" of the journals are less read by men of power — even by those in the mass media, but that "middlebrow" journals or quality journals aiming at a market of some 300,000 or more are definitely on the readers' list of a fair proportion of men of power

TABLE 33

The "Trickle-Down" Effect: Magazines Read by Elites in Mass Media and Other Fields (In Percentages)

MAGAZINE OR JOURNAL	MASS MEDIA LEADERS READ[a]	THE ONE OTHER SECTOR WHOSE LEADERS READ[a] MOST FREQUENTLY	
Newsweek	98	Congress	91
New York Times Magazine	93	Politically appointed administrators	75
Time	88	Congress	91
New Yorker	86	Owners	77
Harper's	85	Congress	50
Fortune	75	Industrial corporations	100
New Republic	75	Labor	40
Atlantic	75	Congress	39
		Politically appointed administrators	39
Saturday Review	64	Congress	55
Foreign Affairs	64	Congress	45
		Politically appointed administrators	45
New York Review of Books	61	Politically appointed administrators	25
Esquire	59	Labor	36
Commentary	54	Congress	25
National Review	47	Congress	41
Commonweal	39	Congress	16
Nation	37	Labor	48
Daedalus	37	Politically appointed administrators	23
Reader's Digest	33	Owners	62
Public Interest	29	Politically appointed administrators	10

[a] At least "occasionally."

and especially form a background to the feelings and opinions of mass media leaders.

Writing and speaking on the war or on any other particular issue was not necessarily an *alternative* to direct influence, as some theorists on the role of intellectuals might have it. Intellectuals with either regular or irregular channels to men of power were *more* likely to speak on the issues or the war – between 75 and 80 percent of them did so – than were others. Part of this may be traced to the Administration's needs to justify itself to intellectual audiences, especially on college campuses, and part of it to the greater visibility of public advisers. For almost all the intellectual elite were *asked* by others to speak; almost never did they volunteer themselves. Much the same held true for writing – most were asked to contribute a piece, but in this case there was little difference between intellectuals with channels to men of power and those without such means of influence.

The most important correlate of writing and speaking was of course intellectual prestige – those with greater prestige were much more likely to write and speak. As we said, this is a two-way street. The greater the prestige, the more intellectuals were likely to be asked to contribute an article or to speak at a meeting. At the same time, the more often an intellectual wrote or spoke, the more likely our respondents were to remember his name and thus to mention him in the course of our many questions about influence and intellectuals.

Given these facts about prestige, we should note that nonacademics were especially active in speaking on the war in Vietnam – 70 percent did so. On non-Vietnam issues it was the high-prestige professors who were most often the "expert" speech makers, though nonacademics whatever their prestige were also quite active. As for writing, social science professors were relatively *in*active on Vietnam, further lending support to the idea that some of them "ducked for cover" when the going got tough. On the other hand, almost every one of them wrote on the issue of his choice.

We have already said that on the Vietnam issue, those with more radical views were more likely to have written, but the point bears frequent repetition, since this fact has led to a distorted view of the intellectual community as a whole. Perhaps because of the need of the intellectual community to find worthy opponents for debate, it appears that those with moderate or conservative positions on the war were more likely than others to have reported that they had spoken on the war.

Conclusion

When I began thinking about this study of elite American intellectuals in 1967, I supposed that the intellectuals had much to do with the labeling and the social recognition of social problems. Social problems, I wrote in proposing this study, are moral problems, and the definition of morality as well as the activation of moral indignation are social processes. Though poverty was hardly created by fiat, social attention to this problem and even its definition have a traceable history from Malthus (1803) to Harrington (1962). Civil rights and the problem of the Negro are not the invention of Negro and white liberal leaders, but it is hardly likely that these issues would share, with poverty, the center of so much attention without some active social efforts. Each one of these issues has, in fact, a traceable history of development in which intellectuals have played leading roles. Recently the Cold War and its "hot" derivatives such as the war in Vietnam have become central to the thinking of many Americans – or so I thought.

Intellectuals have important roles in the process of defining social problems, I said, especially in giving content and form to the value concepts which signify a social problem, for the intelligentsia are increasingly becoming important as opinion leaders on moral issues. Now I am not so sure. The facts are mixed and unclear. The elite American intellectuals, the data suggest, may actually have had relatively little effect on the turn of events of the sixties, though they may have affected the climate of opinion. There is no question but what they were in the vanguard on the main issue of the decade – the war in Southeast Asia. But the war was declared over by Richard Nixon in 1973, some eight years after the majority of the elite American intellectuals came to oppose it. And though the terms for ending the war were negotiated by a leading American intellectual, the terms were such that most of our sample would hardly have approved them and most certainly would have doubts about how well they might work out. Most important, the data show that the majority of American elite intellectuals were not moral leaders on the war; rather, they pointed out its practical difficulties. On the other hand, they were indeed followed in this appraisal some years later by the rest of the American people. Whether intellectuals contributed to this drift or were merely themselves more sensitive to basic American values is hard to say.

The War on Poverty, which I thought still had a chance in 1967, was partly a casualty of the war in Vietnam; by 1972 it had been declared officially dead by President Nixon and his lectuals came to oppose it. And though the terms for ending the tuals in our sample were interested in any kind of social reform and those who were for the most part despaired of success and (as was pointed out above, Pages 256–261) did not know what to propose. Still, on social issues many intellectuals had once been in the vanguard. Another crucial issue, by everyone's account from the public to top American leadership, was civil rights and race relations; at least it was held to be crucial during the first part of the decade until a combination of the war and economic crises crowded it out. Elite intellectuals, though early defenders of civil rights in the 1930's until the Cold War and McCarthyism seemed to make other matters more important, appeared when the big push began anew in the late fifties not as the vanguard, but as followers behind the leadership of Blacks and college students. When the Blacks shucked off white supporters, when urban riots had run out their course because there were no longer rising expectations within the Black community, and when the New York City schoolteachers' strike hopelessly muddled the differences between progressives and reactionaries, the American intellectual elite was left far behind, out of the picture with little to say. In their appraisal of the importance of racial problems, however, intellectuals remain quite far out in front of the rest of the elite and the public, too.

On one issue of the late sixties, an issue which is still with us, the elite intellectuals did clearly lead, but this was an issue which apparently had few direct policy implications. The so-called "culture crisis" represented a reaction on the part of the intellectual leadership to new claims in the realm of values and life-style made by counterculture intellectuals, "new journalists," and college students. This crisis, underlined by the mass media, was associated with the war in Vietnam and the failure of American society to live up to its potential. The leading American intellectuals sharply rejected the "movement" as a culture of lotus-eaters. The nation rejected it, too, in the 1972 elections. After the disclosures of Watergate, however, concern over the nature and structure of American values seems even greater and certainly even more apt. The "paranoid" aura which seemed to surround the Nixon administration, however, makes the "paranoid" feelings of the "movement" a rational response to a difficult situation. From this point of view, and with the benefit of hindsight, the "kids" were right and the intellectual elite, for the most part, were wrong. So in the one area in which they were clearly leaders and clearly the most "expert," most intellectuals failed to draw what in retrospect may seem to have been the right policy implications. But the failure to be right hardly prevents policy advisers from making policy. What the intellectual line on the culture crisis did accomplish, together with their line on Vietnam, was to diminish their political effectiveness by alienating their natural and very large constituency – the college youth.

The content and style of their ideas was not, however, the major factor which prevented elite American intellectuals from having an important effect on American policies in the sixties. Rather, the elite intellectuals were not effective agents mainly because of the style with which they approached other national leaders and the way they saw their own role in society. Which is to say, elite American intellectuals failed to affect policy mainly because they did not really want to take a direct approach, or at least policymaking and advising had a lower priority for most of them than some other matters.

Despite their interest in values, culture, politics and morality, most elite American intellectuals of the kind we have been talking about do not see their major role in society as one of participation in political decision-making or even as regularly advising men of power. Despite this attitude, most American intellectuals do have an *interest* in social policy and are quite active in writing and speaking about it. It is only when they wish *directly* to influence policymakers that most do so in inept, or at least ineffective, ways. As scholars, thinkers, and leaders in their own spheres, they are accustomed to being taken seriously. Moreover, as elites and celebrities in their own right, they receive a good deal of homage. A leading professor is something of a show-business celebrity in his own world and is usually surrounded by a group of admiring persons. Many free-lance intellectuals similarly attract a following. Editors and key journalists are really very powerful people in their own world, with the ability to "make" or "break" others. As a result, when intellectuals offer an idea to men of power they generally prefer (1) to deal with the people right at the top rather than with assistants, legislative aides and the like, whose job it actually is to absorb the kind of information and ideas intellectuals have to offer, and (2) to make their statement and leave. That is, they tend more to one-shot pronouncements and are then somewhat miffed if their golden words are not carefully followed or if they are otherwise rebuffed.

This picture of pronouncement and retreat is of course not true for persons who have chosen to work within government or closely with it, or for those who have set up such groups as the Institute for Policy Studies. As we pointed out, they are a minority within the elite intellectual community, yet they are likely to increase in number. Now, almost one fourth of elite intellectuals are social scientists, many with technical skills valued by government. Some of these are recruited into the kind of advisory positions that men of power feel more comfortable with — that is, they are the kinds of persons more often found on the list of names given by men of power. But these intellectuals are also generalists and have good connections, especially via the social science literary circle with other elite intellectuals. Hence they serve as a bridge between the circles of elite intellectuals and men of power. The role of intermediary is extremely important and likely to increase in importance. It may well be the major way intellectuals can reach out to men of political power. The role has, of course, been formalized by the past several administrations, but this formalization tends to overcentralize the relationship. Since the official intermediary is likely to have a known and strong point of view, intellectuals with other points of view or in other circles tend to be completely left out. This seems somewhat less likely when some key intellectuals play a technical role while at the same time retaining their general interest and their wide contacts.

Elite intellectuals were less reluctant to engage in some forms of political activism as opposed to direct policy advising. The activities in which most were engaged were definitely not of the radical kind. But intellectuals' efforts here seemed to suffer from the same sort of deficiency as their attempts to advise the powerful. They were unlikely to be permanently integrated into an ongoing structure which had close ties to the American political process. Many of the organizations were of an ad hoc nature (the most effective ones were small elite groups aimed at educating key legislators and others about issues of war, peace and disarmament. One of these lasted for over 20 years, but this was an exception). The general disarray of the Democratic Left – the group most of our intellectuals would best fit in with – of course contributed to their lack of organizational effectiveness. The trickle-down effect from the writings of the elite intellectuals in the elite journals probably remains the chief mode of communication, and as a result, of influence that intellectuals exercise. The process is a most uncertain one, operating, as we have seen, chiefly through the mass media and partly through the staff of key legislators and federal administrators. To what extent intellectuals create a climate of opinion and to what extent they merely reflect it remains unknown. The extraordinary convergence between the elite, the intellectuals, and the public in their view of which issues are important may well be caused by the fact that all are exposed to the mass media. On the other hand, we have clearly shown that the mass media leaders are omnivorous readers of intellectual media, and that while the top intellectual stars may utter their pronouncements in these media only after specialists and technicians have had their say, many of these very specialists also publish in intellectual journals. Even the intellectual stars, who may lag behind the "experts," are clearly out in front of most of the mass media and certainly the public. Who leads whom and whether, as some experts suspect, the real social innovators are actually high-level civil servants and the staffs of some legislators is a matter that can ultimately be settled only by a close study of the development of selected issues. But there is a definite possibility that intellectuals do have an indirect effect on policy through the definition of issues and the creation of a climate of discussion. Of course, all this presupposes that the American people, through the medium of their leaders, really would like to listen to intellectuals if the latter could only somehow learn to communicate to them or to assume some sort of role which would make them more palatable to men of power. On the one hand, it is true that an ever-increasing proportion of high school graduates are now entering college (some 70 percent in 1972) and hence are becoming exposed to the work of the intellectuals we have interviewed. The potential audiences for quality journals and for thoughtful books are increasing absolutely and proportionally. The potential following of American intellectuals must be increasing. On the other hand, without invoking the specter of American "anti-intellectualism," it is still possible to claim that the increasing cultural sophistication of the average American is mainly in the direction of increased attention to the relations between means and ends – to technology – rather than in the direction of a reexamination of the ends themselves in any abstract fashion. And this impatience with anything but quick technical solutions seems even more characteristic of the top American leadership – and not only the leadership of the current administration. Whatever intellectuals might do – even the more technically oriented of the general intellectuals – might simply be unappealing and uninteresting to the rest of the American leadership. For the generalist intellectuals we have been concerned with would probably have to turn themselves into bureaucratic technicians to get full hearing – and then they would simply be replaced by another set of generalist thinkers equally unlistened to. The indirect role of intellectuals in affecting policy therefore remains the crucial one.

We must come back to the notion of circles and networks of intellectuals. While it is always modish to talk about, on the one hand, the lonely man of genius, and on the other hand, the integration of intellectuals into society — and we certainly have seen signs of both in our data — almost all the serious ideas advanced by the intellectuals we have talked with have either resulted from interaction with other intellectuals or were at first tested upon the supercritical audience of other elite intellectuals. Without the stimulation of other intellectuals and without the filtering process engendered by discussion with other elite intellectuals, the very life of ideas is threatened.

This means that important ideas about values, culture morality, politics and esthetics — the sort of things we have been concerned with — are generally developed in a setting which is necessarily somewhat isolated from the rest of America and from the action world of men of power, and which is necessarily composed of intellectuals talking mainly to each other. Given the structure of modern large societies, it could hardly be otherwise. However diffuse the network of intellectuals and however removed from the nineteenth-century salon, the network does turn back upon itself and tends to include only persons who at least have similar dispositions, and similar political ideas and values. Most of the intellectual elite must therefore remain like prophets in the wilderness, proclaiming their principles with little assurance that anyone but a fellow intellectual is listening.

16

Conclusion: The Role of the Elite American Intellectuals

Despite or because of the many works on intellectuals, there is no adequate sociological theory of intellectuals or intellectual life. We do not have such a theory either, though in this brief concluding chapter we shall try to list the ingredients of such a theory and locate elite American intellectuals within that framework. Theory building in this field has been marred by an abundance of opinion and moralization, a dearth of facts, and a plethora of parochial definitions. We have tried to combat these flaws in the present work but cannot here supply what is perhaps the most essential ingredient: a full comparative perspective based on well-grounded research. We do have more facts about the American intellectual elite, however, than have ever been assembled before.

One of the peculiarities of theories about intellectuals is that they tend not to be theories at all but exhortations, moralizations and condemnations – tendencies to which we have succumbed on occasion ourselves. Leading sociologists who have written on intellectuals, including Weber, Veblen, and Mannheim of another generation –not to mention Marx, and Lenin, if they may be included – and Shils, Rieff, Lipset, Bell,

Parsons, Coser, Nettl and others of this generation, seem unable to avoid interjecting their own values and wishes into their analyses. (Merton, though often accused of covertly introducing values, has generally avoided this.) This is not the place to review sociological theories of intellectuals in detail. But the crux of the debate seems to lie in proclamations about the proper role of intellectuals – the subject, as we have seen, of much concern among intellectuals themselves. Contact with the intellectuals they study and the fact that the observers are also intellectuals seems to make it impossible for them not to enter into the debates rampant in the intellectual world. The prescriptive tone of the discussion, a tone we have not been able to avoid either, is of course more prevalent in works on intellectuals by non–social scientists who feel a moral stance is rightfully theirs.

An adequate theory of intellectuals should contain an analysis of the general functions and roles of intellectuals in different types of societies, a description of the social types which play these roles and why they fit into them better than others, and an analysis of the recruitment process; these parts of the theory are, of course, much dependent on the definition of intellectual. A further component of a theory of intellectual life should include an analysis of the structure of intellectual life – the way it is organized, how it links into education systems, its relationship to political life, its economic basis and the degree to which it is "elitist" or "populist." Finally, and the very focal point of the debate on intellectuals, the theory must deal with the content of the ideas of intellectuals: for example, the extent to which they represent a clerisy and are defenders of tradition or the extent to which they represent a major force of dissent and change.

In Chapter 1 we saw that definitional problems plague the study of intellectuals; the conclusions of the argument are often embedded in the definition. If the intellectual is defined as "the man of knowledge" or better, an innovator of knowledge,[1] then it naturally follows that a study of intellectuals will be a study of modern knowledge systems and will inevitably deal with colleges, universities, and research institutes as well as with science and technology. Such a study will show intellectuals to be extremely varied in opinion, to have little awareness of themselves as a class, and to contain a strong conservative element as well as a small more radical fringe. Because of their technological interests and their bureaucratic service functions, modern intellectuals will be found to be well integrated into contemporary organizational life, albeit at some loss to their intellectual independence and with some change in their modes of thought.[2]

If intellectuals are defined as basically concerned with dissent,[3] then they will be found to have opinions on the Left and to be politically effective only during certain periods in history.[4]

If, on the other hand, it is felt that intellectuals are basically concerned with values, then it is possible to find intellectuals as strong defenders of tradition, at least under certain circumstances.[5] And if intellectuals are defined mainly in terms of their social position as an "unattached" stratum,[6] then it should come as no surprise that intellectuals are synthesizers who may often attach themselves to the political struggles of various classes.

Our own definition of intellectuals in terms of experts in values who communicate their ideas to others, together with our insistence that intellectuals must certify each other, is of course just as limiting, especially because we concentrate on purveyors of the printed word or those who influence these writers. Thus we have almost no scientists, and no artists, musicians, actors or filmmakers in our study of the American intellectual elite. But this, we argued, is inherent in the American scene itself. The country is so large both in terms of population and in terms of the lack of a central intellectual capital (after all, "only" half the American intellectuals, by our definition, live in the New York region, and the proportion would have been even smaller had we chosen a more inclusive definition of intellectual), that these different systems tend to become more sharply differentiated than they do in many European countries or in developing nations where the relatively small educated class is forced together by a variety of circumstances. After all, the intellectuals we sampled *could* have named many artists, for example, as persons they talked with. Only a few, mainly art critics, did so. Given this very clear limitation on our part, then what *facts* can we contribute to the debate on intellectuals?

First, as to social origin and recruitment.[7] We find Mannheim to be basically correct, at least as far as the American intellectual elite is concerned. They are recruited from a wide range of social classes, and it also seems that they are "an increasingly inclusive area of social life." We also find with Mannheim that intellectuals' class backgrounds tend to be less important in establishing their opinions than the forces of the intellectual community itself. This as a fact is very difficult to absorb for social scientists raised in a tradition which, if not Marxist, has certainly been highly influenced by Marx and with which Mannheim himself is identified. According to Mannheim, class influences the nature of ideas, but it does so to a lesser extent for intellectuals.

> Although they are too differentiated to be regarded as a single class, there is however one unifying sociological bond between all groups of intellectuals, namely, education, which binds them together in a striking way. Participation in a common educational heritage progressively tends to suppress differences of birth, status, profession, and wealth, and to unite the individual educated people on the basis of the education they have received.

Mannheim is of course reluctant to abandon entirely the notion of class and status even for intellectuals, so he hastens to add, "In my opinion nothing could be more wrong than to misinterpret this view and maintain that the class and status ties of the individual disappear completely by virtue of this."[8]

To be sure, we too have found some residue of status ties, less of class ones. For example, Jewish elite American intellectuals tend to be more politically active in indirect political activities, though not in direct policy advising, and tend to have more intellectual prestige. But today they are by and large not more radical than other intellectuals, though there is some evidence that this may have been true in the 1930's and early forties. And one can get absolutely nowhere with class background in trying to explain the position on issues of American elite intellectuals today. For that matter, studies of college students show that for each year in college, social class origin declines as an explanatory variable for opinions and values of all kinds. Once current field is taken into account, social class origin explains relatively little of the opinions of the American professoriate at large, though religion still counts.[9] These points cannot be overstressed, since both laymen and social scientists persist in thinking otherwise.

While the "common educational heritage" may have been binding in the Germany and pre–World War II England with which Mannheim was familiar, it is probably less so in the United States today. Rather, we argued that despite the American geographic dispersion, the very loosely knit social network of intellectuals who communicate with each other and take each other into account is the effective socializing agent for the American intellectual elite. Here we agree with Shils, who notes that the many different intellectual communities – scholarly, scientific, literary, and artistic – express and enforce the basic standards of the intellectual world and are the basic structural building blocks of this world.

> They operate like a common-law system without formal enactment of their rules but by the repeated and incessant application and clarification of the rules. The editors of learned scientific, scholarly and literary journals, the readers of publishing houses, the reviewers of scientific, scholarly and literary works, and the appointments committees which pass judgment on the candidates for posts in universities or scientific research institutes are the central institutions of these communities. . . . Intellectual communities remain really effective systems of action.[10]

Our interests here are more narrow and our gatekeepers have been noted as the intellectual journals, but the point is basically the same. And at least for the kind of intellectual we have been talking about, the editors of the journals are often no more powerful than the social network or circle to which they are for the most part only loosely related. At the time of our study in 1970 there were two major American elite intellectual circles which intersected to form a third, central circle. There was a circle chiefly composed of literary intellectuals, many of whom had become radicalized by the sixties, and larger circle of literary and social science intellectuals, many with a strong leftist background, who were now "liberal" in the sense of the middle fifties. There was also a much smaller and more compact circle of radicals, and a still smaller circle of biologists, scientists and others interested in the environment. Neither the ecology movement nor the radical movements of the sixties managed to penetrate into the mainstream of elite American intellectual life, or at least to affect its basic social structure.

One of the most important characteristics of the social circles of elite American intellectuals is that they tend to represent the resolution of past issues rather than current ones.[11] By and large, the circles we described represent alignments of the late fifties, with some effects of the early sixties. This surprising fact may account for the nonradical (in their own terms) stance of most of the elite American intellectual community in 1970. For despite the rumblings of the sixties there arose no new structures with wide appeal among elite intellectuals which reflected the apparent changes of the decade. The Vietnam issue crosscut the existing networks without, at least in 1970, having changed them. This is a crucial finding to which we shall return, for without an expression in social structure, ideas tend not to be effective or long-lived.

Another important structural fact about elite American intellectuals is the role of the university in American intellectual life. The enormous growth after World War II of American universities focused much attention upon them as the chief institution of American intellectual life. While universities are crucial for scholarly and scientific activity, this is not entirely the case for the elite intellectuals concerned with values.[12] Though it has been claimed that the number of "unattached intellectuals" has been declining, we found that in 1970 15 percent of the American intellectual elite were true free-lancers, and only 40 percent were professors. The rest are editors and staff members of various publications, though a number of professors also held such positions and everyone did some free-lance work. Thus we agreed in detail that neither universities nor other institutions connected with scholarly activities certify the American intellectual elite. Moreover, the networks of intellectuals are *not* related to universities, and this despite the fact that Columbia, Harvard, Yale and New York University account for half the posts of the professors in our study. The state of American universities and the health of the education system generally does affect the world of the intellectual elite, however, in ways we shall discuss below.

No analysis of controversies in theories of the intellectual can take place without considering the matter of economics. There have been a number of attempts to account for the alleged alienation of intellectuals in Europe by their relative lack of employment, income, and integrated position in the economies of their respective countries. But as Ben David has pointed out,[13] this has certainly not been the case for England, Germany or Sweden. Nor is it clear that present-day French or Italian intellectuals, for example, though by and large radical, are especially impoverished or divorced from the mainstream of their economies. As for the intellectuals in our study of the United States, they are clearly members of the economic "establishment," at least the information system part of that establishment – the media and the universities. The median income of the members of our sample who answered the question was $35,000. While much lower an income than that reported by other sectors of the American elite, it is still a respectable figure and indeed is much higher than that reported by a sample of leading American university professors. And it should be noted that this income is generally an earned income (very few of the intellectual elite have private means), which enmeshes the intellectual in the American system in two ways. First, the 85 percent of the intellectuals who were not free-lancers had offices, schedules, meetings, committees, and secretaries. Most of them supervised more than their own incomes and were frequently in charge of substantial budgets. Thus elite American intellectuals are organizationally meshed into America. (So much for the claim by populists that intellectuals have never had to meet a payroll.) Second, the reason the income reported by the intellectual elite was higher than that of an equivalent sample of professors is that the intellectuals obtain a considerable part of their income from royalties, which the professors do not. In many ways, these intellectuals are the last of the traditional entrepreneurs! Though I have not studied the matter in detail, I suspect that both organizational involvement and a high degree of entrepreneurship characterize the life of the intellectual elite of most advanced societies. In eastern Europe, with its system of extra jobs and high payment for scientific or scholarly publication, intellectuals are one of the few entrepreneurial types in the system.

With these basic facts about the position of elite American intellectuals, we can consider the various claims about what their role is or should be in relation to each other and to the rest of the society. Despite their lack of agreement on other matters, Parsons, Shils, Nettl and Coser, among current sociological observers of intellectuals, seem to agree that a basic element of the role of the kind of intellectual we are talking about here is the analysis, development, revision, representation and even sometimes the creation of basic values and opinion. Whatever the relative stress on those aspects of the role, it does constitute their primary influence upon modern society. Thus Parsons noted, "Ideology has become the primary instrument of the modern secular intellectual classes in their bid to be considered generally important. . . . This is of course a matter of generalized influence on the affairs of society."[14] Shils talks about the primary role of intellectuals in working with patterns or symbols of general significance and notes that in Western societies prior to the twentieth century, intellectuals' influence upon society was mainly through the creation of patterns of belief. Indeed, even in the present century, one of their major impacts has been the creation of both modern constitutional politics and the doctrine for revolutionary movements. At the same time, both Parsons and Shils observe a vast increase in what Shils calls the secondary role of intellectuals: that is, the performance of practical activities in which intellectual works are involved, or what he calls "intellectual executive actions."[15] This technical function, we have seen, has important implications even for the kind of intellectuals with which we are dealing and is a matter we shall return to shortly. In any case, even Nettl, considerably to the left of Parsons and Shils, agrees that intellectuals are self-consciously experts on "basic" culture, and are concerned with "universals" rather than "particulars."[16] But both he and Coser add the notion that intellectuals are also basically social-political dissenters;[17] this has to do with the content of the ideas they espouse, not the level at which they are operating.

Despite the implication or the direct claim that intellectuals must not only create and interpret what Shils calls "symbols of general significance"[18] but disseminate them as well, we find little examination among sociologists of the mechanisms and devices through which this crucial filiation of ideas takes place. Somehow, the assumption is made that what we have called "indirect influence" does take place but that the details are not overly important.[19]

The chief method chosen by contemporary intellectuals for the dissemination of their ideas is the intellectual journal, so the world of such journals must be carefully analyzed if the role of intellectuals in the dissemination of new ideas is to be properly understood. We showed that despite or perhaps because of their relatively small circulation, intellectual journals have the power to make or to break the reputation of intellectuals. One of the more curious findings, and one we shall return to, is that the leading intellectuals of the moment tend *not* to initiate ideas and topics of controversy but to wait until some of the first smoke has settled and *then* to offer their assessments of the situation. To be sure, this process of having the foot soldiers advance before the generals does lead to an occasional rapid promotion of those experts and initiators who are able to show that they are more than one-piece or one-theme writers.

But the world of intellectuals' journals is rapidly changing and where it will go is hard to say. The late sixties saw the *New York Review of Books* jump far into the lead of the other journals as it presented a more radical view of the war in Southeast Asia and the problem of American Blacks. With the seventies, however, the *Review* has dropped some of its more strident reviewers, *Commentary* has turned from a centrist-liberal position to a more conservative one,* and *Partisan Review* continues on its liberal-left course but with diminished influence. New journals, such as *Public Interest* and *Modern Occasions* (the latter already defunct), have appeared. The intellectual journals with wider circulation such as *Harper's* and *Saturday Review* have experienced sharp twists and turns. The

* Though the editor still feels the journal maintains a liberal view.

New York Times Book Review has a new editor for the seventies, one considerably more modish and Left than the older man who had retired after twenty-one years of editorship. The *New York Times Magazine,* widely read by both intellectuals and men of power, has also acquired new direction with a new editor of the *Sunday Times.* The Left or underground newspapers of the sixties, with the exception of the *Rolling Stone,* have mainly failed. The most unsettling aspect of the situation with intellectuals' journals, however, is the absence of clear circles and networks around most of them. Since the early part of the nineteenth century, intellectual life in Great Britain, to a certain degree in France, and later in the United States, has mainly centered around networks, circles and salons attached or related to particular intellectual journals. The journals had a "line" and the intellectuals associated with it expounded it — albeit with variation and dissent. Now this is not so clear in the United States. The lines are not as clearly drawn as some observers of the Podhoretz-Epstein rift would have it. Whether the current loose structure is the result of an inexorable press of mass society, the growing absorption of intellectuals into the geographically dispersed university structure, which tends to create circles based more on discipline than on general interests, or simply a transition phase is not possible for us to say at the moment.

The argument that American intellectuals are now in transition has much to recommend it. To sense these changes we need more of an historical perspective. In the brief historical aside which follows, we draw on the late Richard Hofstadter, a liberal, and the currently very active Irving Howe, a democratic socialist. Both men were very important themselves in the events of the sixties which we have recounted in the previous chapters.

Hofstadter begins at the beginning of the Republic, and his views are succinctly summarized by his title to the last chapter of *Anti-Intellectualism in American Life* — "The Intellectual: Alienation and Conformity." He argues that American intellectuals have from the very founding of the country oscillated between alienation and conformity. "In our earlier days," he notes, "two groups of intellectuals were associated with or responsible for the exercise of far-reaching social power, the Puritan clergy and the Founding Fathers." But this association with power was replaced in the nineteenth century by the "mugwump mind," which retained the respect for reason and order of the Founding Fathers. But "having been edged out of the management of [the country's] central institutions . . . the patrician class produced a culture that became overrefined, desiccated, aloof, snobbish . . . the genteel tradition. Its leaders cared more that intellect be respectable than it be creative."[20] Their conformity to an intellectual tradition served to alienate them from populist America. Perhaps not the least of their alienation was their almost total inability to earn a living at their intellectual pursuits. If the nation rejected them as effeminate and ineffectual, they themselves were "quite ambivalent about their America," complaining about it and yet at the same time identifying with it.[21]

In the earlier part of the nineteenth century in America there were perhaps no encompassing or relatively stable circles in intellectuals, in our terms. Hofstadter observes, however, that "after 1890 it became possible for the first time to speak of intellectuals as a class" or a community.[22] A number of intellectual institutions were developed, including major universities, great libraries, important journals, and so on, not to mention the development of Bohemian communities – all institutions which encouraged the growth of what we have been calling intellectual networks or circles. But precisely as "American writers and other intellectuals became a more cohesive class . . . they took up arms against American society," and "alienation became a kind of fixed principle among knowing young intellectuals during the years preceding the war [World War II]."[23] A growing group self-consciousness and cohesiveness which is combined with a sense of differentness from other groups is, of course, characteristic of the social psychology of group life, and if the combination is "ironic," as Hofstadter would have it, it is merely the irony of life itself. Self-awareness and group cohesiveness also tend to emerge when groups become more powerful and more important to the national society. The direction which group culture may take is determined by historic circumstance, but that the result is an extreme one is hardly surprising when a group has suddenly come of age. Be that as it may, by the twenties, American intellectuals "seemed to outdo each other to prove that there was no such [American] civilization."[24]

Just as the situation in the United States became economically intolerable in the 1930's, American intellectuals began another oscillation in the direction of pro-American feeling which ended in the apparent total conformity of the early fifties.[25] The New Deal, the new labor movement, the sense of a new era together with the evident growing sickness of Europe, made American innocence and morality much more attractive.

> In 1947, the year when America came to Europe's rescue with the Marshall plan, Edmund Wilson, the least provincial of writers [and the recipient of the most nominations in our study for prestige among intellectuals], found it possible to say upon returning from Europe that "the United States at the present time is politically more advanced than any other part of the world," and to speak of our twentieth century culture as a "revival of the democratic creativeness which presided at the birth of the republic and flourished up to the Civil War." The twentieth century, he felt, had brought "a remarkable renascence of American arts and letters."[26]

As Conway pointed out in his chapter on the Cold War in this book, and as Howe demonstrated in his "The New York Intellectuals,"[27] reactions to Stalinism and to the Cold War were complex and multistranded. "It is absurd, and indeed disreputable," said Howe in 1968, "for intellectuals in the '60's to write as if there were a unified 'anti-communism' which can be used to enclose the views of everyone from William Buckley to Michael Harrington."[28] Howe, nonetheless, an editor of *Dissent,* added,

> For a hard line group within the American Committee for Cultural Freedom, all that seemed to matter . . . was a sour hatred of the Stalinists. . . . The dangers in such a politics now seem all too obvious, but . . . in the early 50's they were already being pointed out by a mostly unheeded minority of intellectuals around *Dissent.* . . . The criticism to be launched against the New York intellectuals [mainly, our center circle] in the post war years is not that they were

strongly anti-Communist but rather that many of them . . . allowed their anti-Communism to become something cheap and illiberal.[29]

What is more, most American intellectuals came to believe that "The America . . . of vast inequalities and dramatic (the quote is from Mary McCarthy) contrasts is rapidly ceasing to exist," and, felt Howe, "the New York writers all but surrendered their critical perspective on American society."[30]

According to Howe, the *Partisan Review* "failed to speak out with enough force and persistence" against McCarthyism, and *Commentary,* "under Elliot Cohen's editorship, was still more inclined to minimize the threat of McCarthyism." Yet our data show that by the late fifties, most of the intellectuals in our sample, including the "New York Intellectuals," had become soured on many of the basic concepts of the Cold War.

Finally, in gaining perspective on the ideas of the American intellectual elite in the years prior to our study, we should add Howe's observations on the inclinations, or lack of them, of the New York intellectuals toward politics (again, we are extrapolating to the intellectual elite as a whole, but not unfairly):

> I have been speaking here as if the New York intellectuals were mainly political people, but in reality this was true for only a few of them. . . . Most were literary men or journalists with no experience in any political movement; they had come to radical politics through the pressures of conscience and a flair for the dramatic; and even in later years, when they abandoned any direct political involvement, they would in some sense remain "political." . . . The real contribution of the New York writers [however] was toward creating a new . . . style of work. . . . They were continuing a cultural movement that had begun in the United States during the mid-19th century: the return to Europe, not as provincials . . . but as equals in an enterprise which by its very nature had to be international.[31]

That is, the American intellectual elite were both creating new works and creating important criticism that was to link them firmly, and in a new way, to "the modern masters." The embeddedness of elite American intellectuals in the European and Western tradition of culture, whatever their flirtation with mass culture, cannot be overstressed if one is to understand their reaction to the events of the sixties.

Thus the ground is set for understanding the political and cultural ideas of the American intellectual elite during the sixties and the incomplete changes which it produced. The degree to which intellectuals represent an "adversary culture," to use Trilling's apt and oft-quoted phrase, the degree to which they are more of a clerisy than a revolutionary force, their linkages to modern technocratic society, and their actual role in that society — that is, the issues at the very core of the debate about intellectuals — seem to me to be dependent in large measure on a correct assessment of the *content* of the ideas of intellectuals in different societies.

The most informed current debate on the role of American intellectuals places them in the context of intellectuals in other industrially advanced countries.[32] Lipset and his colleague Dobson claim that American, as well as Soviet Union, intellectuals are clearly dissenters and that this creates problems for a society so dependent on intellectual functioning. Jill Conway (no relation to Thomas Conway), on the contrary, says American intellectuals "are adversaries about means and ends in realizing the American dream, not about its ultimate desirability." Malia grants that while American intellectuals may be critical, this is by no means necessarily a consequence of their intellectuality and, contrary to the claims of Lipset and Dobson, intellectuals in the Soviet Union operate within an extremely narrow critical horizon. "Both the institutional and the cultural context which they operate make their criticism far less sweeping and global than that of their more fortunate Western confreres."

What then have we found is the true content of the political and social ideas of the American intellectual elite of the sixties? What relationship does this content have to the role of elite intellectuals and their influence on policies? At the level of opinion-poll data, we found in 1970 the American intellectual elite to be to the left of any other sector of the American elite and even to the left of comparable American university professors. Thus there is no question but what the general climate of opinion of elite American intellectuals is one of opposition and generalized dissent on such catch phrase issues as withdrawal from Vietnam, marijuana, government controls of industry, civil rights, student uprisings and the like. This finding is in line with what might be expected of elite American intellectuals; they are not mid-America and even the more conservative among them when confronted with the simple choices of a public opinion poll find themselves on the opposite side of the fence from most Americans.

There is more to the ideas of intellectuals, however, than opinion-poll responses (though on many issues there is very little more to the ideas of most Americans). Because it was the crucial issue of the decade, we spent a good part of this book documenting orientations of elite intellectuals to foreign policy in general and to Southeast Asia in particular. It is true that most American intellectual elite began to change their ideas on the Cold War after the death of Stalin, even before Kennedy promised to get America moving again. They were probably the first major group in American society, perhaps even before the college students, to oppose the war. Certainly they were in advance of the average liberal college professor. The very structure of the intellectual media seemed organized to report and to ferret out goings-on that do not fit with official views. And the network of intellectuals worked very well indeed to spread the word — and opposing views too — to every corner of the elite intellectual community. The fact that so many intellectuals are now in universities helped too, since most of the laggards were brought along by the college student mood. All this certainly amounts to dissent.

In assessing the role of the intellectual elite in opposing the war in Vietnam we stressed three points, however, that mitigate this impression of dissent. First, the most generally prestigious intellectuals appeared in print only *after* less prestigious intellectuals who were specialists in Southeast Asia or in foreign affairs had had their say. Timing is of crucial importance in understanding the role of the intellectual elite and we shall return to this point. Second, the residue of the American self-satisfaction and pro–Cold War attitudes of the fifties, as well as the pragmatic traditions of America, led most of the intellectual elite to oppose the war not because it represented innate imperialistic forces in the United States, not because it was morally wrong, but simply because it did not work. Those morally opposed and especially those ideologically opposed tended to get more space in print, but their views were not representative of the majority of the intellectual elite.

Our findings on reasons for opposing the war, when fed back to some of our respondents in 1973, seemed to rankle them. With the hindsight of 1973, many (except those most on the Left, who said they suspected this all along) did not want to believe that the intellectual elite reasoned like the rest of the American people, though the elite were much quicker to see the point. They would like to have believed that in the sixties intellectuals, at least, saw the moral implications if not the full structural implications now revealed by Watergate. But the printed record, as well as their 1970 interviews, simply does not back up this hindsight. Perhaps in the excitement of mass student rallies on campuses during the Cambodian invasion in the spring of 1970 many waxed moralistic. But these very same persons, when calmly reviewing the war in our interviews conducted around the same time, were pragmatic rather than moralistic in tone. And although blatant intervention was thought to be wrong, the intellectuals had not yet, in 1970, evolved a new "line" on foreign policy.

The final point on Vietnam was the general impotence of the intellectual elite to do much about their views. There were some important members of the intellectual community who held key jobs in the Kennedy administration and some even lasted well into the Johnson administration. Most of these favored intervention in Vietnam but later came to regret their views. By then, neither the counsel of former experts and advisors nor the entreaties of other intellectuals who did have channels to Washington were able to reverse the situation. The Nixon administration which followed did employ two of the better-known members of the American intellectual elite as domestic and foreign policy advisors. This meant that almost any intellectual who wished to could contact them. But Moynihan played it very cagey on the war, and in any case claimed to have no impact on foreign policy, while Kissinger was a leader of the 20 percent or so of the intellectual elite who took a hard and pragmatic line on foreign policy. "Henry" was available to most of the intellectual elite who wanted to talk with him, but he obviously was not influenced, which gave the intellectuals no choice but to utilize indirect means of influence, including, of course, their best weapon, writing.

Continuing the heritage begun by the Civil Rights movement, many elite intellectuals took to the picket line and engaged in marches and other demonstrations, though some of the more active ones complained that many social democrats became activists only late in the game. Whatever they did was, however, ineffective. Public opinion seems to have been swung more by the events in Vietnam than any actions taken by intellectuals. Perhaps it is too much to expect that the intellectual elite should have appreciable impact on the course of the nation's policy. Yet their potential constituency in the 1960's was proportionately and absolutely greater than it had ever been before, perhaps because of the youthful population bulge proportionately greater than it ever will be again. Nonetheless, they were not able to reach that constituency in any appreciable degree.

Given the turn of cultural events in the sixties, it is not certain that they could in fact have reached college and university youth. Those intellectuals deeply interested in foreign policy either clung to a *realpolitik* point of view with little appeal to the young or attempted to maintain a general interest in peace, disarmament and world government, an interest they had held since the end of World War II, but an interest which was eclipsed by the war in Vietnam. Then too, the sixties were an era of mass movements, and the peace–antibomb circle among elite intellectuals was more attuned to giving seminars to key senators than to staging mass rallies.

Finally, even had the youth been captured by the moral arguments of the leading intellectuals, it is far from certain that the rest of America would have followed. Certainly, sensitivity to moral issues was not evidenced by the public in the Nixon-McGovern campaign.

The Civil Rights movement represented another debacle for elite intellectuals. In the first place, though race relations had been a long-term interest for American intellectuals and one of the major foci for their dissent from America, McCarthyism and the Cold War diverted attention from this issue, which was raised with force again only in the late fifties and early sixties and then not by the intellectual elite but by Black leaders and college youth. Most intellectuals were happy to follow – but it was clearly a case of following and not leading. The final blow came when the Blacks turned against white leaders in their quest for independence and self-emancipation. And since 99 percent of the American intellectual elite in the sixties was white, this meant that they were effectively cut out of the Civil Rights movement. Those intellectuals still deeply interested in the Civil Rights movement in 1970 had in any case been discouraged and had almost totally run out of ideas.

Except among radical Marxist intellectuals, who persistently saw a connection between foreign policy and domestic social reform, elite intellectuals' interests in general social equality, educational reform and other domestic matters took second place in the late sixties to the consuming matter of the war. This again was a change over which the intellectuals had little control, for just as they were beginning to recover from a consuming interest in anti-Communism and were becoming again influenced by those who had steadfastly remained democratic socialists, the war in Southeast Asia demanded full attention. The result was again, as with the Civil Rights movement, that the few intellectuals concerned with socialism or domestic reform for the most part, even from their own point of view, were at a loss for what to propose to revitalize the forces fighting for social equality and justice.

But the most serious problem for the American intellectual elite during the sixties was in some ways not the war but what we have called "the great culture crisis." Both Hofstadter and Howe, though their politics were quite different, end their essays reviewing the history of the American intellectual elite with a full-scale attack on the anti-intellectualism and cultural naïveté of beats, the counterculture, the so-called New Left, and new journalism and the writings of Mailer (a true member of their circle, however). Even a strong self-professed radical, an ex-disciple of Marcuse, academic David Kettler, vigorously attacks the Marcuse of *One-Dimensional Man* and deplores the abandonment by young radicals of serious theorizing.[33] Very few of the intellectuals in our sample had much good to say for the counterculture and the general intellectual style of the so-called New Left.[34] Their lack of regard for the traditions of Western thought was more than almost any of the elite intellectuals could bear, even those regarded as heroes by the young.

Thus, between their pragmatic stand on the war, their disdain for the thought patterns of the college and graduate student generations of the sixties, and their inability to come up with new ideas on domestic reform, the American intellectual elite were caught in a very difficult situation, saved only by the fact that most did not realize how serious their predicament was. The intellectual elite did reject the dominant culture and were indeed far to the left of the American people. Yet they were outflanked on the left by the first generation since the 1930's that might have been ready to accept their ideas. They would not and could not lead the "far-out" (ranging from Andy Warhol to Tom Hayden) with their alien life-styles and non-logical modes of thought. The result has been at least one, perhaps two "lost generations" of youth who have not had some leading members coopted into the intellectual elite. This has had at least two important consequences. One we have already alluded to: that is, the diminished political effectiveness of the intellectual elite by virtue of their having lost a large ready-made following. The second has been an incomplete transition from the Cold War days to some new position. For transitions to eventuate into new ideas requires new social forms, which often follow as a result of new personnel. The circles of American intellectuals have not basically changed, in part because there have been few additions to them during the sixties, and the loss of these generations means that replacements are still not in evidence.

This is the way it was in the sixies and it is unreasonable to fix "blame" upon either the intellectual elite or the radical youth for having missed what seems to me to have been a great opportunity to transform America. Even in retrospect it seems doubtful that given their history, the forces from outside acting upon them — the war, the assassinations, and "paranoia" of the ruling class which eventuated in Watergate (a "paranoia" matched only by the views of the radical youth), and the long-term trend of mass culture which in rock and roll finally attracted the "best and the brightest" among the youth – that either the youth or the elite intellectuals could have acted in any other way. Still, one wishes the intellectuals had been quicker to oppose the war on moral or ideological grounds and that the youth had been more receptive to logic and history.

What role then did the American intellectual elite play in the sixties and what are they likely to do in the seventies? First, it appears that they were not originators of new moral ideas but rather analysts, critics and disseminators – crucial tasks, but many intellectuals might prefer to think of themselves as originators. Whether this nonoriginating role is generally true of elite intellectuals – and one can make a good case for this – or is an artifact of this decade in the United States is not something I could argue at this point. Second, they are dissenters indeed, but still a loyal opposition. Their lack of nihilistic radicalism, in contrast with the youthful radicalism of the sixties, is the source of the debate between those who see them as a clerisy and those who see them as critics, dissenters or rebels. Not only are there divisions within their ranks – a minority of elite intellectuals in America are indeed cultural and/or political radicals – but the tendencies toward clerisy and dissent are often wrapped up within the same person. Thus their role in the sixties was to oppose and to propose – but generally nothing too far removed from liberal sensibilities. And as for the creation of new values, this simply did not occur. Whether intellectuals far removed from the "people" can ever create new values is another matter, but one which cannot be settled without good comparative data.

In any case, what has been described is at least in attitude suspiciously like an "Establishment." Irving Howe claims that such a term is at best an overstatement when applied to the American intellectual elite. He contrasts the so-called English Establishment to the situation in the United States.

> "Establishment" does not bear the conspiratorial overtones. . . . What it does in England is to locate the social-cultural stratum guiding the tastes of the classes in power and thereby crucially affecting

the tastes of the country as a whole. In this sense, neither the New York writers nor any other group can be said to comprise an American Establishment, simply because no one in this country has ever commanded an equivalent amount of cultural power. The New York writers have few, if any, connections with a stable class of upper-rank civil servants or with a significant segment of the rich. They are notably without connections in Washington. They do not share official or dominant tastes. And they cannot exert the kind of control over cultural opinion that the London Establishment is said to have maintained until recently. . . . They [the New York writers] lack – and it is just as well – the first requirement for an Establishment: that firm sense of internal discipline which enables it to impose its values on a large public.[35]

Thus Howe suggests that whatever they think and whatever their values, the intellectual elite hardly affect American culture and if they cannot influence the goings-on within their own domain it seems even less likely that they can do so in politics. This point of view is distinctly in opposition to that of Parsons and Shils, who think intellectuals do have generalized cultural influence, at the very least. Our own findings lie somewhere between these views. True, elite American intellectuals, whether in New York or elsewhere, have relatively few direct connections to men of power – but they have more than Howe thinks they do. What ties there are exist mainly because of a trend obscured by Howe's focus on the literary elite. A significant number of the intellectual elite are, as we pointed out, social scientists who play the dual role of "mental technicians" (H. Stuart Hughes's brilliant phrase) and elite intellectuals. As such they do have relations to power and serve as conduits for other intellectuals not so inclined.[36] Whether this relationship to power destroys their intellectuality or not is hard to say, but despite the present national administration it seems difficult to believe that the trend of some intellectuals acting as advisors to men of power and even taking on decision-making roles will not continue.

Parsons, Shils and Lipset all take a broader view of the term "intellectual" than Howe or myself and include all the mental technicians, university professors, scientists and the like.[37] As such, to say that they do not have a general influence on the course of a society's ideas, values and politics would be absurd. And with the growth of higher education this influence must become greater. But the issue for Howe and myself is whether or not a core group of elite intellectuals concerned with values has such an influence. Here the findings are again somewhat equivocal. If one judges by results, then Howe may be right, though it must be said that the present political and cultural mood of the country far better fits the mood of the intellectual elite in 1970 than it does either the mood of the apparently revolutionary sixties or the mood of Backlash America. On the other hand, data reported in the previous chapter suggest that at least the rest of America's elite are more exposed to the writings of the intellectual elite than the intellectuals think, and this exposure is especially heavy among the mass media elite.

The role of elite American intellectuals has therefore been in the sixties to oppose both youth and the powerful, but not totally, and to communicate this opposition for the most part indirectly, to men of power and via the mass media to the rest of the country. The opposition has not been with a single voice, but it has been more uniform than one might think. While there were some positive proposals in the early sixties, the intellectuals found little to propose to the country in the late sixties.

In an unsettled situation prognoses are the most difficult to make, though it is then, of course, when they are most in demand. We have mentioned several symptoms of instability: the lack of new ideas (though the resurgence of a debate on equality seems especially promising); the lack of new recruits; a loosening of the bonds of the social networks without the appearance of a new set of structures; the uncertain situation of the intellectual journals; the almost total divorce from the Nixon administration; and a halt in the expansion of universities, an expansion which provided jobs for many of the intellectual elite and especially for whatever protégés they may have had.

Most commentators and observers of elite intellectuals worry about alienation. Too much is a problem — after all, it is most uncomfortable. Too little is equally a problem — in the modern world at least, too much embeddedness in the routine structural life of a nation seems to impede intellectual creativity. Despite their apparent alienation in the sixties, most of the elite American intellectuals were simply too well off and too enmeshed in the daily routines of bureaucratic life to qualify as highly alienated and this embeddedness may have affected their style of thought.

The handwriting may be on the wall, but in script too small to notice. The irony may well be that after a touted decade of apparent discontent and radical feeling – neither of which was much reflected in the American intellectual elite – a new era of alienation may set in, accompanied by new social forms, new circles and new journals.

Appendix

Sampling the American Intellectual Elite

Julie Hover

The process of obtaining a sample of elite American intellectuals to be interviewed required four steps. Beginning with the assumption that influential intellectuals would write in, or have books reviewed in journals that were influential among intellectuals, a study was undertaken to identify these journals.[1] Next, a content analysis was carried out on journals selected by four or more persons as being influential among intellectuals plus two journals (*Public Interest* and *Foreign Affairs*) frequently written-in by respondents on a subsequent study of journals. One out of every four journals issued weekly, one out of every two journals published monthly, and every issue of journals published quarterly were analyzed for the years 1964–1968. This process resulted in a list of approximately 20,000 entries, with each entry representing an article, column, poem, book review, or book reviewed. About 8,000 *persons* were in this list, 61 percent of whom appeared but once, 18 percent but twice, 6 percent three times and 15 percent four or more times. A mere .7 percent appeared more than twenty-five times.

By limiting the sample to persons who had at least four entries (or three entries including at least one book), approximately 1300 *persons* comprised the universe from which the sample was drawn. The sample was biased in favor of persons who wrote most frequently. This was accomplished by sampling *entries* (rather than persons) by random number. The more entries a person had, the greater the probability that one of his entries would be randomly selected. The success of the self-weighting technique is evident in the following figures.

NUMBER OF ENTRIES PER PERSON	PERCENTAGE IN LIST OF THOSE ELIGIBLE	PERCENTAGE IN SAMPLE ACTUALLY DRAWN
3–5	51	17.5
6–10	30	35.0
11–20	12	25.0
over 20	7	22.5

In addition to being weighted, the sample was stratified. The first stratification was by type of entry, with three strata (article writers, book reviewers, and authors) included. Within each of these strata, journals were grouped according to prestige, influence, and content (political, literary, and mixed). Thus the sample was stratified to represent type of entry and type of journal with each stratum weighted according to the number of entries per author.

Sample of Writers of Articles and Fiction

In order to draw the sample of article and fiction writers, the journals were divided into six groups. The first five groups of journals were as follows:

(1) *Ramparts* and the *New York Review of Books*
(2) *Commentary, Dissent, Partisan Review* and *Public Interest*
(3) *Harper's, Atlantic Monthly, Daedalus* and *Reporter*
(4) *Nation* and *New Republic*
(5) *Kenyon Review, Yale Review, Saturday Review, New Yorker* and the *New York Times Book Review.*

Forty persons were selected from the five groups of journals in the following way. First we totaled the number of articles for all five groups (column 2) and determined what percentage of these articles appeared in each group (column 3). Then we took that percentage of forty. (Column 3 x 40 persons equals column 4.) For example, journals in group one comprised 6.44 percent of the total number of articles in all five groups. Since 6.44 percent of forty is three, 3 persons were randomly selected from the 198 writers of articles and fiction in group one. This process resulted in figures as follows:

JOURNAL GROUP	NUMBER OF CONTRIBUTORS[2] IN GROUP	NUMBER OF ARTICLES WRITTEN BY CONTRIBUTORS	PERCENTAGE OF TOTAL NUMBER OF ARTICLES	COLUMN 3 x 40	NUMBER ACTUALLY SAMPLED
1	61	198	6.44	3	3
2	159	913	29.72	12	12
3	153	607	19.76	8	9
4	97	431	14.03	5	4
5	211	922	30.02	12	12
					40

Since Group 4 included only two journals, only 4 persons (rather than 5) were sampled, while 9 persons (rather than 8) were included in Group 3. In addition to these 40 persons, 4 names were randomly selected from each of the five groups, and 11 persons were randomly selected from this list of 20 persons. A total of 51 persons were sampled from these five groups of journals.

The sixth group of journals, judged to be less influential, included *American Scholar, Commonweal, Esquire, Paris Review, Progressive, Sewanee Review,* and *Foreign Affairs.* One person was randomly selected from the list of 1184 entries (235 persons) included in this group of journals. A total of 52 writers of articles and fiction was included in the sample of persons to be interviewed.

Sample of Book Reviewers

Four persons were included from journals judged to be influential about literary matters among intellectuals. Three persons were randomly selected from the 1067 entries (of 191 reviewers) in the *New York Review of Books, Commentary,* and *Partisan Review.* One person was randomly selected from the 1196 entries (225 reviewers) appearing in the *New York Times Book Review* and *Saturday Review.*

Authors of Books Reviewed

To draw the sample of authors, journals were divided into four groups. The first three groups were:

(1) *Commentary, Partisan Review* and the *New York Review of Books*
(2) The *New York Times Book Review* and *Saturday Review*
(3) *Atlantic Monthly, Daedalus, Dissent, Harper's, Kenyon Review, Nation, New Republic, New Yorker, Reporter, Yale Review* and *Public Interest.*

Since each group included approximately the same number (approximately 1000 entries or 275 authors in each group), seven persons were randomly selected from each group. Three authors were selected from the 678 entries (219 authors) reviewed in the journals considered to be less influential (listed above). Thus 24 persons who had books reviewed were included in the sample.

Editors[3]	22
Article and Fiction Writers	52
Book Reviewers	4
Authors of Books Reviewed	24
Total	102

Replacements

Persons in the sample who would not or could not be interviewed were replaced through randomly selecting persons with the same qualifications as those imposed on the person they replaced. Editors, however, were not replaced. An additional 50 were drawn for this "reserve," making a total draw from our master list of 152. Of those on this list we managed to interview 97, a very high rate, we think, in view of the nature of the sample.

The fourth and final step was the "snowball." In order to make sure we obtained the list of leading intellectuals as defined by themselves, about one third of the way through the interviews, or after the thirty-fourth interview, we tallied all the names mentioned in answer to any of our sociometric questions. A total of 36 who were on our master list of 8000 contributors to journals was mentioned eight or more times. All but 3 were on the eligible list of 1300. Of these, 21 were already drawn on the list of 152 by the very mechanical procedure just described. Clearly, we were on the right track. We added to the list of people to be interviewed the 15 we did not already have. Not wanting to be entirely parochial and wishing somewhat to enlarge our scope, we looked for persons who were *not* on our master list of journal contributors. We had to go down to three or four mentions before we obtained anyone! We added 5 more names in this fashion, all of whom were mentioned three or more times. Thus a total of 20 names was added to the 152 already drawn. Of this "snowball" sample, we managed to interview 13 – a somewhat lower ratio than for the rest, though we had less time to try to arrange meetings with them. The entire sample, then, consisted of 97 persons drawn through our sampling procedures, of whom 12 were also "snowball" candidates, and 13 additional "snowball" persons not already drawn, for a total of 110.

The difference between the 110 interviewed and the 62 not interviewed was tested for every background factor available through directories, as well as the number of articles and books on our master list and the final tally of sociometric mentions. The only significant difference (which affected a few other differences related to it) was our poor record with free-lancers, whose underrepresentation just reaches the .05 level on a Chi Square test.

Notes

CHAPTER 1

1. It is impossible in a work of this kind systematically to review this literature. The bibliography in the back contains some selected references.
2. Paraphrased from Michael Confino, "On Intellectuals and Intellectual Traditions in Eighteenth and Nineteenth Century Russia," *Daedalus,* 101 (Spring, 1972):118. This piece is the best I have found on the subject.
3. Frédéric Bon and Michel-Antoine Burnier, *Les Nouveaux Intellectuels* (Paris: Editions Cujas, 1966), reproduction of original text of the "Manifeste" on page 83.
4. For a study of signatories of these petitions, see Everett Carl Ladd, "Professors and Political Petitions," *Science,* 163 (March 28, 1969): 1425–1430.
5. T. R. Fyvel, *Intellectuals Today: Problems in a Changing Society* (New York: Schocken, 1968).
6. Robert K. Merton, "The Intellectual in Public Bureaucracy," *Social Theory and Social Structure* (New York: Free Press, 1967).
7. Lewis Coser, *Men of Ideas* (New York: Free Press, 1965), p. x.
8. Seymour Martin Lipset, *Political Man* (New York: Doubleday, 1960), p. 311.
9. Talcott Parsons, " 'The Intellectual': A Special Role Category," in *On Intellectuals: Theoretical, Case Studies,* ed. Philip Rieff (New York: Doubleday, 1969), p. 4. In a book published after this was written, Parsons and Gerald Platt, *The American University* (Cambridge, Mass.: Harvard University Press, 1973), define "Intellectual" more narrowly, clearly differentiating the role from that of academic. Their use of the term and their analysis in their Chapter 6, "The University and the 'Intellectuals,' " is consistent with ours.
10. J. P. Nettl, "Ideas, Intellectuals, and Structures of Dissent," in *On Intellectuals,* ed. Rieff, p. 81.
11. Max Kadushin, *The Rabbinic Mind* (New York: Blaisdell, 1965), pp. 1–7, 107–111.
12. Kadushin, *Rabbinic Mind,* pp. 111–112. Significant symbols may be traditional ones. See Juan Linz, "Intellectual Roles in Sixteenth- and Seventeenth-Century Spain," *Daedalus,* Vol. 101 (Summer, 1972), pp. 59–108, for an analysis of "intellectuals identified with an order, religious or political."
13. Sir Charles Percy Snow, *The Two Cultures and the Scientific Revolution* (New York: Cambridge University Press, 1959), esp. "Intellectuals as Natural Luddites." See also the discussion by Snow and others in the *Times Literary Supplement,* October 25, 1963.
14. Parsons does not follow his own definition strictly, for he does include the sciences in his discussion of intellectuals, though in his book *The American University,* scientists are seen to play a different role from "intellectuals."
15. From one point of view, certification is the name of the entire game

in science. For a sophisticated analysis of the many facets of the recognition of excellence in science see Harriet Zuckerman, "Stratification in American Science," *Sociological Inquiry,* 40 (1970):235–257; *Scientific Elite: Studies of Nobel Laureates in the United States* (Chicago: University of Chicago Press, 1972); and Robert K. Merton, "Patterns of Evaluation in Science: Institutionalization, Structure and Functions of the Referee System," *Minerva,* 9 (January, 1971):66–100.

16. For a more extended discussion of the theory and method of social circles and networks see Charles Kadushin, "The Friends and Supporters of Psychotherapy: On Social Circles in Urban Life," *American Sociological Review,* 31 (December, 1966):786–802; and "Power, Influence and Social Circles: A New Methodology for Studying Opinion-Makers," *American Sociological Review,* 33 (1968):685–699.
17. Stanley Milgram, "The Small-World Problem," *Psychology Today,* 1 (1967):1. Jeffrey Travers and Stanley Milgram, "An Experimental Study of the Small-World Problem," *Sociometry,* 32 (December, 1969): 425–443.
18. For an informal history of social circles of various types, see Coser, *Men of Ideas.*
19. The minimum estimate (around 1863) is of 3,000 recognized male painters in Paris and another 1,000 in the provinces. Women painters, occasional painters and professional artists in other fields who did some painting were excluded. See Cynthia and Harrison White, *Canvases and Careers* (New York: Wiley, 1965), p. 83.
20. Coser, *Men of Ideas,* p. 75.
21. Ibid., p. 77.
22. Joseph Ben David and Awram Zloczower, "Universities and Academic Systems in Modern Societies," *Archives Européennes de Sociologie,* 3 (1967):45–85.
23. Our special thanks to Everett Carl Ladd and Seymour Martin Lipset, who made this data available to us for special tabulations. All interpretations are our own.
24. Some of this bias may reflect the time lags in nominations and the tendency of these organizations to nominate yesterday's stars rather than the contemporary ones. Jews became more prominent among elite American intellectuals in the late 1930's and early 1940's. Some of the bias is also caused by yesterday's admission standards. The membership list is the cumulative result of many years' nominations and the requirement of some organizations that there be a lag between the proposal of a name and the eventual election. This tendency is clearly visible in the Century Club's list when examined by year of entry.

    Complete technical details of the sampling methods are in the Appendix.
25. The ins and outs of intellectual journals are described in Charles Kadushin, Julie Hover, and Monique Tichy, "How and Where to Find Intellectual Elite in the United States," *Public Opinion Quarterly,* 35 (Spring, 1971):1–18; and in Chapter 2.
26. For a general discussion of "Jewish Academics in the United States," see the article by Seymour Martin Lipset and Everett Carl Ladd, by that title, in the *American Jewish Yearbook,* 1971.
27. This is shown by the Bureau of Applied Social Research's study of

American leaders directed by Allen Barton, Bogdan Denitch, Carol Weiss, and myself, of which this study is a part.

28. See also Ladd and Lipset in the June 9, 1972, issue of *Science*.
29. Among the American professors generally, both generation and religion were correlated with liberalism on most issues.
30. This list certified by intellectual institutions and circles is different from a list of celebrity intellectuals, as noted by Lewis Coser, "The Intellectual as Celebrity," *Dissent*, 20 (Winter, 1973), pp. 46–56.
31. If we count *all* questions asking for names, other than those on Vietnam issues, then half received fewer than two "votes," 20 percent between two and five, and 30 percent six or more.
32. Most of the writers associated with this revolution – the "new journalists," avant-garde art and film critics, style-setters like Andy Warhol, etc. – have made little if any contact with the intellectuals we have identified in this study, and therefore were less likely to be mentioned by our interviewees.

CHAPTER 2

1. A similar confessional is Mary Breasted's "What New York Does to Its Writers," *Village Voice*, June 1, 1972.
2. Norman Podhoretz, *Making It* (New York: Random House, 1967), p. 351.
3. Kadushin, Hover, and Tichy, "How and Where to Find Intellectual Elite."
4. McLuhan does, however, and it should be noted that he had made "solid" contributions before his celebrity status was gained. For a brilliant analysis of the celebrity problem and journals see Lewis Coser, "The Intellectual as Celebrity."
5. In sampling journals, we included each issue of quarterly journals. For journals published monthly, odd-numbered months were coded for 1964, 1966, and 1968 and even-numbered months for 1965 and 1967. For weeklies, we sampled every fourth issue beginning with the first issue published in 1964, the second in 1965, the third in 1966, the fourth in 1967 and the first in 1968.
6. The famous cover was published on the August 24, 1966, issue. For a discussion of reactions and the way the "Molotov cocktail" got on the cover, see Philip Nobile, "A Review of the *New York Review of Books*," *Esquire*, 77 (April 1972), p. 107.
7. To be sure, Nobile, "A Review," notes that the *New York Review of Books* has now dropped many of its leftist writers named as offensive by our sample.
8. See Page 234.
9. Joseph Epstein, "The New Conservatives: Intellectuals in Retreat," *Dissent*, 20 (Spring, 1973):153. Epstein names with qualifications the following as "new conservatives," and all have published in *Commentary* or *Public Interest:* Edward Banfield, Daniel Bell, James Coleman, Nathan Glazer, Morris Janowitz, Harry Johnson, Irving Kristol, Seymour Martin Lipset, Daniel Patrick Moynihan, Norman Podhoretz, and

James Q. Wilson. All, it turns out, except for editors Kristol and Podhoretz, are social scientists – which implies that a technical orientation is a characteristic of conservatism as Epstein sees it. But there is also a movement in thought and literature, led by such persons as Lionel Trilling, which might be called "reasserted liberalism," which focuses less on technical political criticism and more on values, philosophy and literature. For example, a recent issue of the *New York Review of Books* published a charming essay by Alfred Kazin, a regular contributor, on "Melville the New Yorker," hardly an exercise in "swinging sensibilities," but then the *New York Review of Books* never really embraced that style of literary criticism. Its radicalism was political, not cultural.

CHAPTER 3

1. For an influential view on New York circles by a major participant see Irving Howe, "The New York Intellectuals: A Chronicle and a Critique," *Commentary*, 46 (October, 1968):29–51. See also his debate with Irving Kristol, *Commentary*, 47 (January, 1969).
2. Contrary to the expectations of the sociology of knowledge, their views are not systematically related to their own location.
3. See Michael Harrington for some "White Horse" stories, *Esquire* (August, 1972).
4. Edward A. Shils, *Intellectuals and the Powers* (Chicago: University of Chicago Press, 1972), p. 157, suggests that this envy has been a characteristic of American intellectuals at least since the middle of the nineteenth century.
5. See, for example, Philip Nobile, "A Review of the *New York Review of Books*," *Esquire*, 77 (April, 1972):103–126; Merle Miller, "Why Norman and Jason Aren't Talking," *New York Times Magazine*, March 26, 1972; Dwight Macdonald, "Norman Cousins' Midcult 'World,'" *Columbia Forum*, 1 (Fall, 1972); Dennis H. Wrong, "The Case of the *New York Review*," *Commentary*, 50 (November, 1970):49–63; Peter Steinfels, "The Cooling of the Intellectuals," *Commonweal*, 94 (May 21, 1971):255–261.
6. For background see David Rogers, *110 Livingston Street* (New York: Random House, 1968); Nathan Glazer, "Blacks, Jews, and the Intellectuals," *Commentary*, 47 (April, 1969):33–39; Martin P. Mayer, *The Teachers' Strike: New York: 1968* (New York: Harper, 1969); Barbara Carter, *Pickets, Parents and Power: The Story Behind the New York City Teachers' Strike* (New York: Citation Press, 1971).
7. See comments to this effect in Nobile's interviews, "A Review of *The New York Review of Books*," p. 209.
8. The computational and programming breakthrough was achieved by Richard Alba, under a National Science Foundation grant to the author. The mapping program is called SOCK, and is described in *Behavioral Science*, 17 (May, 1972):326–327. The computer is necessary because there is no reasonable way of locating the points by hand so that the map can be read.

9. There are some other problems with this assumption, but they do not crucially affect findings in this kind of "open system" data.
10. The lines encompassing the circles are not drawn arbitrarily. Rather, they are the result of a computer program called COMPLT, developed by Richard Alba, the basic mathematics of which is described in his "A Graph-theoretic Definition of a Sociometric Clique," *Journal of Mathematic Sociology,* 3 (March, 1973):113–126. Basically, COMPLT identifies cliques according to the criterion that all the individuals within a given clique must be within some fixed social distance of each other. The social distance between two individuals is equivalent to the number of intermediaries (plus 1) required to connect them; that is, if one person is required to connect two individuals, then the distance between them is two. The maximum possible social distance between any two members of a clique must be decided upon in order to use the program. In the case of the cliques drawn onto Figure 1, connections are allowed via at most one other person: that is, the maximum permissible distance is two. Note that both this distance and the general concept of a clique used here follow from our notions of social circles as described in Chapter 1. In the case of the cliques drawn onto Figure 1, this procedure produced 208 cliques. Obviously, they must be overlapping, with many persons in more than one clique. The program next groups all cliques which differ by only one member, providing that relations between them remain sufficiently dense, that is, more dense by a statistically significant amount than the average density of the entire map. This step produced 10 cliques, quite a reduction from 208! But these cliques still overlapped to a great degree. Most of the cliques overlapped by at least 40 percent of the membership of the smaller clique, and some by as much as 96 percent. The program then further clusters the cliques provided that their density remains statistically significant. This resulted in the five cliques shown on the map, two of which are completely nonoverlapping. (The *Partisan Review* clique has one member in common with the social science and literary intellectual clique, but this connection is not shown on the map.)
11. The size represents some persons in our sample and some outside the sample, including some who were not on the original list of 172 but who were brought into the clique by at least two persons within it. Thus there are several biologists in this clique, though there was but one on the list of 172. This must be considered as an estimate of the relative size of the clique, not its absolute number. When data on opinion and other such matters is brought in, it generally represents the number of persons we interviewed or who were on our master list of 172 and includes the "peripherals": that is, persons closely linked to a clique by only one step but not fully part of it. Occupational data, residence and religion are based on my own recognition of the names of all the members and all the peripherals.

    For a view of French academic circles, see Terry and Priscilla P. Clark, "Le Patron et son cercle: Clef de l'université française," *Revue Française de Sociologie,* 12 (1971):19–39.
12. These findings should also be compared with those of Irving Howe, "The New York Intellectuals: A Chronicle and a Critique," *Commentary,* 4 (October, 1968):29–51. Howe notes that "the kind of inner

fraternity associated with literary groups in Paris and London has rarely been characteristic of American intellectual life." But "the New York intellectuals are perhaps the only group America has ever had that could be described as an *intelligentsia.*" It is, however, in its own mind, "a loose and unacknowledged tribe." Now that it is perceived as a group, the "publicity signals recognition and recognition is a certificate of death." The death of the circle is prematurely bemoaned by Howe, and the data he describes fits best with what we have called the center circle, 70 percent of whom are in New York but only three fifths of whom are Jews, in contrast to Howe's statement, "The intellectuals of New York who began to appear in the 30's, most of whom were Jewish." His remarks about looseness and a sense on the part of the members that they are *not* a group fit our data quite well.

13. These figures are slightly different from those reported in my article in the *Public Interest.* The present figures are based on all the persons on the map associated with these circles, rather than sample members only as reported in the *Public Interest.* The assignment of religion is based on my own impression, rather than on the respondent's data, as supplied there. My own assignment takes into account the general family background even when respondents themselves claim to be second-generation atheists!
14. This shows that the issue taken *alone* was supported by many persons who in 1968 supported the teachers' union.

CHAPTER 4

Journals were a basic source for this study. The following were consulted for the period 1947–1962: *The American Mercury, The American Scholar, The Antioch Review, The Atlantic Monthly, The Bulletin of the Atomic Scientists, The Christian Century, Commentary, Common Cause, The Commonweal, Dissent, Foreign Affairs, The Freeman, Harper's, Human Events, Liberation, Masses and Mainstream, Modern Age, The Monthly Review, The Nation, The National Review, The New Leader, New Masses, The New Republic, Partisan Review, The Progressive, The Reporter, The Saturday Review of Literature, Science and Society, The Virginia Quarterly Review,* and *The Yale Review.*

1. See Robert A. Divine, *Second Chance: The Triumph of Internationalism in America during World War II* (New York: Atheneum, 1967). Divine is especially useful for information on the efforts of such UN supporters as Clark Eichelberger and James T. Shotwell. On the Atlantic and World Federalists see Clarence Streit, *Union Now* (New York: Harper, 1939), and his journal, *Freedom and Union* (1946–); also see Ely Culbertson, *Summary of the World Federation Plan,* sponsored by the World Federation, Inc. (New York: Garden City Publishing Co., 1943), *Total Peace* (New York: Doubleday, Doran, 1943), and *Must We Fight Russia?* (Philadelphia: Winston, 1946). On the World Constitutionalists see the journal *Common Cause* (1947–1951), edited by G. A. Borgese.
2. For example, see Norman Cousins, *Modern Man Is Obsolete* (New

York: Viking, 1945), and his editorials in *Saturday Review;* for a collection of his best editorials see *Present Tense: An American Editor's Odyssey* (New York: McGraw-Hill, 1967).

3. On the scientists' movement see Alice Kimball Smith, *A Peril and a Hope: The Scientists' Movement in America, 1945–1947* (Chicago: University of Chicago Press, 1965). Also see the *Bulletin of the Atomic Scientists,* which was founded in 1945 and edited by H. H. Goldsmith and Eugene Rabinowitch (edited by Rabinowitch alone after 1949); besides contributions from such scientists as Karl Compton, Albert Einstein, I. I. Rabi, Leo Szilard, Edward Teller and Harold Urey, the journal also published the work of such nonscientists as Abba Lerner, Charles Merriam, Talcott Parsons, Bertrand Russell, Edward Shils, James T. Shotwell, and Quincy Wright, particularly during its early and most active years.
4. For Lippmann's wartime views see his *U.S. Foreign Policy: Shield of the Republic* (Boston: Little, Brown, 1943), and his *U.S. War Aims* (Boston: Little, Brown, 1944); also see Francine C. Cary, *The Influence of War on Walter Lippmann* (Madison, Wis.: State Historical Society, 1967), pp. 139–174, and Edward L. and Frederick H. Schapsmeier, *Walter Lippmann: Philosopher-Journalist* (Washington, D.C.: Public Affairs Press, 1969), pp. 83–104. For Spykman's views see his *America's Strategy in World Politics: The United States and the Balance of Power* (New York: Harcourt, Brace, 1942).
5. On the anti-Stalinist liberals see Clifton Brock, *The Americans for Democratic Action: Its Role in National Politics* (Washington, D.C.: Public Affairs Press, 1962), and the ADA newsletter, *The ADA World.* For the views of a major spokesman of postwar neo-liberalism, Reinhold Niebuhr, see his journal, *Christianity and Crisis,* and such books as *The Children of Light and the Children of Darkness* (New York: Scribner's, 1944), *Discerning the Signs of the Times: Sermons for Today and Tomorrow* (New York: Scribner's, 1946), and *Christian Realism and Political Problems* (New York: Scribner's, 1953). For the statement which best summarizes the anti-Stalinist liberal position see Arthur Schlesinger, Jr., *The Vital Center* (Boston: Houghton Mifflin, 1949). A good study of the development of liberal values in the early postwar years is Walter Christian Schuman's "Structures of Liberal Thought in America, 1945–1952" (unpublished honors thesis, Harvard University, 1969).
6. On these anti-Stalinists, who made New York City their center, see Irving Howe, "The New York Intellectuals: A Chronicle and a Critique," *Commentary,* 46 (October, 1968):29–51; Irving Kristol's criticism and Howe's reply can be found in *Commentary,* 47 (January, 1969):12, 14, 16; also see Norman Podhoretz, *Making It* (New York: Random House, 1967). In the late 1940's many of these intellectuals shed what remained of their Marxist and socialist pasts. See, for example, the symposium, "The Future of Socialism," *Partisan Review,* vol. 14, nos. 1–5 (1947), and the symposium, "Our Country and Our Culture," *Partisan Review,* vol. 19, nos. 3–5 (1952). On this latter symposium see Richard Hofstadter, *Anti-Intellectualism in American Life* (New York: Knopf, 1962), pp. 394–398. For a novelistic treatment of this theme see Lionel Trilling, *The Middle of the Journey*

(New York: Viking, 1947). An extended treatment of the *Partisan Review* circle in the 1930's and 1940's can be found in James Burkhart Gilbert's *Writers and Partisans: A History of Literary Radicalism in America* (New York: Wiley, 1968), pp. 118–232.

7. The nascent conservatives differed from the liberal and leftist anti-Stalinist intellectuals in the 1940's primarily because the former were beginning to identify freedom and democracy with free enterprise and were generally more bellicose in their attitudes toward both the Soviet Union and domestic communists.
8. The contributors to *Politics* (1944–1949) included Nicola Chiaromonte, Lewis Coser (Clair), Paul Goodman, Peter Gutman (Meyer), C. Wright Mills, Victor Serge and Niccola Tucci. A collection of Macdonald's best essays from the 1940's can be found in his *Memoirs of a Revolutionist* (New York: Farrar, Straus and Cudahy, 1957). For more on Macdonald see Gilbert, *Writers and Partisans,* and Christopher Lasch, *The New Radicalism in America, 1889–1963: The Intellectual as a Social Type* (New York: Knopf, 1965), pp. 322–334. On Norman Thomas and the Socialists see Harry Fleischman, *Norman Thomas: A Biography, 1884–1968* (New York: Norton, 1964), and Norman Seidler, *Norman Thomas: Respectable Rebel* (Syracuse, N.Y.: Syracuse University Press, 1961). For an examination of the peace movement during the 1940's see Lawrence Wittner, *Rebels Against War: The American Peace Movement, 1941–1960* (New York: Columbia University Press, 1969); for information on peace activist A. J. Muste see Nat Hentoff, *Peace Agitator: The Story of A. J. Muste* (New York: Macmillan, 1963).
9. *Masses and Mainstream,* a monthly, first appeared in March, 1948; it replaced both the *New Masses,* which had ceased publication in January, 1948, and the Marxist literary quarterly, *Mainstream,* which had begun publication in 1947.
10. On the Wallace movement and its intellectual supporters see Curtis MacDougall, *Gideon's Army,* 3 vols. (New York: Marzani and Munsell, 1965), particularly I, 102–127, and Karl M. Schmidt, *Henry A. Wallace: Quixotic Crusade, 1948* (Syracuse, N.Y.: Syracuse University Press, 1960). For a hostile portrait of Wallace see Dwight Macdonald, *Henry Wallace: The Man and the Myth* (New York: Vanguard, 1948), and for an interesting description of Wallace's tenure as editor of the *New Republic* see Bruce Bliven, *Five Million Words Later: An Autobiography* (New York: John Day, 1970), pp. 265–274.

    Journals sympathetic to Wallace, such as the *New Republic* and the *Nation,* finally did not endorse him, primarily because of their fear that his third party would split the liberal vote. On the communists' support of Wallace see Joseph Starobin, *American Communism in Crisis, 1943–1957* (Cambridge, Mass.: Harvard University Press, 1972), pp. 155–194.
11. For example, the *Progressive* kept its leftist credentials by endorsing Thomas.
12. On the fragmentation of the scientists' movement see Edward Shils, "Freedom and Influence: Observations on the Scientists' Movement in the United States," in *The Intellectuals and the Powers and Other Essays* (Chicago: University of Chicago Press, 1972), pp. 154–195.

On the disintegration of the peace and internationalists' movements see Wittner, *Rebels Against War,* pp. 182–239. Spokesmen such as Clarence Streit and Norman Cousins reluctantly came down on the side of United States policy; others lapsed into silence. Dwight Macdonald stopped editing *Politics* in 1949 and turned his energies to criticism of mass culture. For a critical analysis of those intellectuals who turned from social-political criticism to criticism of mass culture in the late 1940's and 1950's see Edward Shils, "Daydreams and Nightmares: Reflections on the Criticism of Mass Culture," *The Intellectuals and the Powers,* pp. 229–247. Criticism of Cold War policies did continue; in fact, in 1949 Paul Sweezy and Leo Huberman began publishing the radical-socialist journal, *The Modern Monthly.*

13. For one of the few strongly dissenting views on the Korean War see I. F. Stone, *The Hidden History of the Korean War* (New York: Monthly Review Press, 1952). For an analysis of his position see Norman Kaner, "I. F. Stone and the Korean War," in *Cold War Critics: Alternatives to American Foreign Policy in the Truman Years,* ed. Thomas G. Paterson (Chicago: Quadrangle Books, 1971), pp. 240–265.
14. On this concept, so crucial to much Cold War political thought, see Hannah Arendt, *The Origins of Totalitarianism* (New York: Harcourt, Brace, 1951), *Totalitarianism,* ed. Carl J. Friedrich (Cambridge, Mass.: Harvard University Press, 1954), Carl J. Friedrich, Michael Curtis and Benjamin R. Barber, *Totalitarianism in Perspective* (New York: Praeger, 1969), and Carl J. Friedrich and Zbigniew Brzezinski, *Totalitarian Dictatorship and Autocracy* (Cambridge, Mass.: Harvard University Press, 1956).
15. The ACCF, one of the most prestigious organizations of intellectuals in the early 1950's, with membership by invitation only, was established in 1951. Members in the early 1950's included Hannah Arendt, Max Ascoli, James Baldwin, Roger Baldwin, William Barrett, Jacques Barzun, Arnold Beichman, Daniel Bell, Saul Bellow, Robert Bendiner, Eric Bentley, Paul Bixler, McGeorge Bundy, John Chamberlain, Whitaker Chambers, Richard Chase, Leo Cherne, Elliot Cohen, George Counts, Norman Cousins, David Dallin, Robert Gorham Davis, Moshe Decter, Ralph de Toledano, Theodosius Dobzhansky, John Dos Passos, F. W. Dupee, Max Eastman, Merle Fainsod, James T. Farrell, Carl Friedrich, Harry Gideonse, Nathan Glazer, Clement Greenberg, Oscar Handlin, Sidney Hertzberg, Granville Hicks, Eric Hoffer, Richard Hofstadter, Sidney Hook, Quincy Howe, Oscar Jaszi, Horace Kallen, Murray Kempton, Hans Kohn, Irving Kristol, Abba Lerner, Saul Levitas, Seymour Martin Lipset, Arthur Lovejoy, Russell Lynes, Dwight Macdonald, Ernest Nagel, Allen Nevins, Reinhold Niebuhr, J. Robert Oppenheimer, William Phillips, Philip Rahv, A. H. Raskin, David Riesman, James Rorty, Richard Rovere, Arthur Schlesinger, Jr., Robert Shaplen, George Shuster, Upton Sinclair, Mark Starr, John Steinbeck, Fritz Stern, Allen Tate, Norman Thomas, Frank Trager, Diana Trilling, Lionel Trilling, Peter Viereck, Robert Penn Warren, James Wechsler, Alan Westin, Morton White, William L. White, Thornton Wilder, and Bertram Wolfe. It should be noted that there were factions of opinion on the Cold War within the liberal and

leftist membership of the Committee. Not all of them agreed with the Stalwart approach, but the general thrust of the organization was in support of it.

One of the major themes to emerge from both the ACCF and its parent organization, the Congress for Cultural Freedom, was "the end of ideology"; for a collection of many of the documents relevant to this theme see *The End of Ideology Debate*, ed. Chaim Waxman (New York: Funk and Wagnalls, 1968).

In 1953 the CCF established, in England, the journal *Encounter*, which was edited by Irving Kristol and Stephen Spender until 1958, and then by Melvin Lasky and Stephen Spender. Among the American intellectuals who contributed to it were Daniel Bell, Daniel Boorstin, Leslie Fiedler, Nathan Glazer, Seymour Martin Lipset, Dwight Macdonald, Harold Rosenberg, and Lionel Trilling.

For a decidedly unflattering portrait of the CCF and the ACCF see Christopher Lasch, "The Cultural Cold War: A Short History of the Congress for Cultural Freedom," in *The Agony of the American Left* (New York: Knopf, 1969), pp. 63–114.

Journals which advocated the liberal-leftist Stalwart approach (not all contributors agreed with this advocacy) in the early 1950's included:

The New Leader, edited by Saul Levitas. Contributors included Daniel Bell, William Henry Chamberlin, David Dallin, Moshe Decter, Denis Healey, Granville Hicks, Sidney Hook, George Kennan, Reinhold Niebuhr, Stefan Possony, James Rorty, Robert Strausz-Hupe, Diana Trilling, and Bertram Wolfe.

And to a lesser extent:

Commentary, when edited by Elliot Cohen (1946–1959), had among its contributors Hannah Arendt, Daniel Bell, Robert Gorham Davis, Louis Fischer, Nathan Glazer, Clement Greenberg, Oscar Handlin, Sidney Hertzberg, Sidney Hook, Irving Kristol, George Lichtheim, Robert Lekachman, James Rorty, Arthur Schlesinger, Jr., and Bertram Wolfe.

Partisan Review, edited by William Phillips and Philip Rahv, had such contributors as Hannah Arendt, James Baldwin, William Barrett, James Burnham, Richard Chase, Robert Gorham Davis, Leslie Fiedler, Sidney Hook, Irving Howe, George Lichtheim (G. L. Arnold), Arthur Schlesinger, Jr., Diana Trilling, and Lionel Trilling.

The Commonweal, edited by Edward Skillen. Contributors included Philip Burnham, John Cogley, John Lukacs, James O'Gara, George Shuster, and Peter Viereck.

The Christian Century, whose major figures were Charles Clayton Morrison and Harold Fey.

The conservative Stalwart approach was advocated in the following journals:

Human Events (1944–1957), edited by Frank C. Hanighen and Felix Morley, which had such contributors as William F. Buckley, Jr., William Henry Chamberlin, Frank Chodorov, and David Dallin.

The American Mercury. When edited by Lawrence Spivak (1944–1950) and William Bradford Huie (1950–1953), it had among its contributors James Burnham, John Chamberlain, David Dallin, Ralph de Toledano, Max Eastman, Granville Hicks, Sidney Hook, Eugene Lyons, Freda Utley, Peter Viereck, and Bertram Wolfe.

The Freeman (1950–1955), edited successively by John Chamberlain and Henry Hazlitt, Henry Hazlitt alone, Kurt Lasen, and Frank Chodorov. The contributors included Louis Bromfield, James Burnham, William F. Buckley, Jr., William Henry Chamberlin, Forrest Davis, Ralph de Toledano, Isaac Don Levine, Milton Friedman, Victor Lasky, Eugene Lyons, Jonathan Mitchell, Raymond Moley, Stefan Possony, Victor Riesel, William S. Schlamm, George Sokolsky, and Freda Utley.

The National Review, edited by William F. Buckley, Jr., since its inception in 1955, had such contributors as L. Brent Bozell, James Burnham, Frank Chodorov, Forrest Davis, Max Eastman, Willmoore Kendall, Russell Kirk, Suzanne LaFollette, Eugene Lyons, Jonathan Mitchell, Frank S. Meyer, Morris Ryskind, William S. Schlamm, and Freda Utley.

Modern Age, a quarterly, was founded in 1957; during the Cold War period it was edited first by Russell Kirk and then by John Davidson. Among its contributors were Harry Elmer Barnes, William Henry Chamberlin, David Dallin, Ludwig Freund, Will Herberg, Willmoore Kendall, John Lukacs, Frank S. Meyer, Thomas Molnar, Felix Morley, Gerhart Niemeyer, Stefan Possony, Ernest van den Haag, Peter Viereck and Richard Weaver.

On the development of conservative thought during the 1950's see William J. Newman, *The Futilitarian Society* (New York: George Braziller, 1961), and Jeffrey Hart, *The American Dissent: A Decade of Modern Conservatism* (Garden City, New York: Doubleday, 1966).

16. See William F. Buckley, Jr., and L. Brent Bozell, *McCarthy and His Enemies* (Chicago: Henry Regnery, 1954).

17. For a liberal-leftist Stalwart view of domestic communism, loyalty programs, and McCarthyism see Sidney Hook, *Heresy, Yes – Conspiracy, No* (New York: John Day, 1953); on McCarthy himself see James Rorty and Moshe Decter, *McCarthy and the Communists* (Boston: Beacon Press, 1954). Also see *The New American Right,* ed. Daniel Bell (New York: Criterion Books, 1955), Dennis Wrong, "Theories of McCarthyism: A Survey," *Dissent,* 1 (Autumn, 1954): 385–392, and Michael Paul Rogin, *The Intellectuals and McCarthy: The Radical Specter* (Cambridge, Mass.: MIT Press, 1967).

18. See Hart, *The American Dissent,* pp. 127–168; also James Burnham, *The Coming Defeat of Communism* (New York: John Day, 1950) and his *Containment or Liberation?* (New York: John Day, 1953), William Henry Chamberlin, *Beyond Containment* (Chicago: Henry Regnery, 1953), Stefan Possony, *A Century of Conflict: Communist Techniques of World Revolution* (Chicago: Henry Regnery, 1953), Robert Strausz-Hupe, *The Zone of Indifference* (New York: Putnam, 1952), Robert Strausz-Hupe, William R. Kintner, James E. Dougherty, and Alvin Cottrell, *Protracted Conflict* (New York: Harper, 1959), and Robert Strausz-Hupe, William R. Kintner, and Stefan Possony, *A Forward Strategy for America* (New York: Harper, 1961).

19. The academic-intellectual journals published a limited variety of views of the Cold War, views which generally fell within the Stalwart approach. If an author was critical, he was critically supportive and his reservations were framed within the language of *realpolitik*. Examples of such journals (again, not all contributors shared the same viewpoint):

The American Scholar, edited by Hiram Haydn. Contributors included William Barrett, Daniel Boorstin, Van Wyck Brooks, Elmer Davis, Robert Gorham Davis, Irwin Edman, Robert Heilbroner, Granville Hicks, Joseph Wood Krutch, Max Lerner, Archibald MacLeish, Saul Padover, David Riesman, Clinton Rossiter, Richard Rovere, and Mark Van Doren.

The Antioch Review, edited by Paul Bixler. Among its contributors were Jacques Barzun, Daniel Bell, William Carlton, Stuart Chase, Theodosius Dobzhansky, Heinz Eulau, James T. Farrell, C. Hartley Grattan, Granville Hicks, Sidney Hook, Alex Inkeles, Horace Kallen, Sidney Lens, David Riesman, James Rorty, and Peter Viereck.

The Virginia Quarterly Review, edited by Charlotte Kohler, had such contributors as Crane Brinton, Carl Dreher, Herbert Feis, Calvin Hoover, Owen Lattimore, Richard Lauterbach, Max Lerner, Ernest K. Lindley, Reinhold Niebuhr, Eugene Rostow, Vincent Sheean, and George E. Taylor.

The Yale Review, edited by Helen McAfee and later David Potter, had among its contributors Frederick Barghoorn, John Chamberlain, David Dallin, Vera Micheles Dean, Herbert Feis, Calvin Hoover, Russell Kirk, Henry Kissinger, Hans Kohn, Sidney Lens, Reinhold Niebuhr, Clinton Rossiter, Robert Strausz-Hupe, George E. Taylor, and Peter Viereck.

The journal published by the Council on Foreign Relations, *Foreign Affairs,* shared the academic-intellectual journals' narrow range of opinion. It was edited by Hamilton Fish Armstrong and had such contributors as Joseph Barnes, McGeorge Bundy, Isaac Deutscher, John Foster Dulles, Bernard Fall, Herbert Feis, Albert Guerard, Henry Kissinger, William Langer, Walter Millis, Philip Moseley, Reinhold Niebuhr, J. Robert Oppenheimer, James Reston, Vincent Sheean, Sumner Welles, Bertram Wolfe and Henry Wriston.

20. For Kennan's views see his *American Diplomacy, 1900–1950* (Chicago: Chicago University Press, 1951), *Russia, the Atom and the West* (New York: Harper, 1958), and *Russia and the West Under Lenin and Stalin* (Boston: Little, Brown, 1961). Also see his *Memoirs, 1925–1950* (Boston: Little, Brown, 1967), especially pp. 354–367, which deal with the famous "X" article, and *Memoirs, 1950–1963* (Boston: Little, Brown, 1972). For Lippmann's views see his *The Cold War: A Study in U.S. Foreign Policy* (New York: Harper, 1947) and *The Communist World and Ours* (Boston: Little, Brown, 1959). For an analysis of his views during the Cold War see Barton Bernstein, "Walter Lippmann and the Early Cold War," in Paterson, *Cold War Critics,* pp. 18–53, and Edward L. and Frederick H. Schapsmeier, *Walter Lippmann,* pp. 105–127. Also see *Walter Lippmann and His Times,* ed. Marquis Childs and James Reston (New York: Harcourt, Brace, 1959),

and Anwar Syed, *Walter Lippmann's Philosophy of International Politics* (Philadelphia: University of Pennsylvania Press, 1963). Morgenthau's views can be found in his *Peace, Security and the United Nations* (Chicago: University of Chicago Press, 1946), *Scientific Man vs. Power Politics* (Chicago: University Press, 1946), *Politics Among Nations: The Struggle for Power and Peace* (New York: Alfred A. Knopf, 1948), *In Defense of the National Interest* (New York: Alfred A. Knopf, 1951), *The Purpose of American Politics* (New York: Alfred A. Knopf, 1960), and *Politics in the Twentieth Century,* 3 vols. (Chicago: University of Chicago Press, 1962). For Niebuhr's views see *The Structure of Nations and Empires* (New York: Scribner's, 1959); for more information on Niebuhr see *Reinhold Niebuhr: His Religious, Social and Political Thought,* ed. Charles W. Kegley and Robert W. Bretall (New York: Macmillan, 1956) and June Bingham, *The Courage to Change: An Introduction to the Life and Thought of Reinhold Niebuhr* (New York: Scribner's, 1961). During the Cold War the *New Republic* had as its editors Henry Wallace (1946–1947), Michael Straight (1948–1956), and Gilbert Harrison (1956–). Its contributors included Louis Fischer, Waldo Frank, Helen Fuller, Louis Halle, Murray Kempton, George Kennan, Leon Keyserling, James King, Jr., Joseph Lash, Max Lerner, Hans Morgenthau, Lewis Mumford, Reinhold Niebuhr, John Roche, Richard Rovere, and Arthur Schlesinger, Jr.

The Reporter, edited by Max Ascoli (1949–1968). Its contributors included A. A. Berle, Robert Bendiner, Robert Bingham, Stuart Chase, Marquis Childs, Theodore Draper, Louis Fischer, John Kenneth Galbraith, Paul Jacobs, Alfred Kazin, Henry Kissinger, Sidney Lens, Walter Millis, Norman Podhoretz, Eric Sevareid, William V. Shannon, Edgar Snow, Edmund Taylor, and James P. Warburg.

21. *Dissent,* edited by Irving Howe and Lewis Coser had such writers as Isaac Deutscher, Erich Fromm, Manny Geltman, Paul Goodman, Michael Harrington, Norman Mailer, C. Wright Mills, A. J. Muste, Stanley Plastrik, Bernard Rosenberg, Ben Seligman and Dennis Wrong. Also see Irving Howe, *Steady Work: Essays in the Politics of Democratic Radicalism, 1953–1966* (New York: Harcourt, Brace and World, 1966).
22. See Carey McWilliams, *Witch-Hunt: The Revival of Heresy* (Boston: Little, Brown, 1950), Fred Cook, *The Warfare State* (New York: Macmillan, 1962), and C. Wright Mills, *The Power Elite* (New York: Oxford University Press, 1956). For responses to Mills's work see *C. Wright Mills and the Power Elite,* ed. G. William Domhoff and Hoyt Ballard (Boston: Beacon Press, 1968).
23. The *Progressive* was reorganized as a new journal in 1948 with Morris Rubin remaining as editor. Among its contributors during the Cold War were Roger Baldwin, Robert Bendiner, McAlister Coleman, Stuart Chase, Vera Micheles Dean, John Haynes Holmes, Sidney Lens, Murray Kempton, Leon Keyserling, Milton Mayer, Richard Rovere, and Norman Thomas. *Liberation* was established in 1956. Its editors were David Dellinger, A. J. Muste, and Bayard Rustin. Among its contributors from 1956 to 1962 were Dorothy Day, Waldo Frank, Michael

Harrington, Paul Jacobs, Sidney Lens, Norman Mailer, Milton Mayer, Lewis Mumford, Robert Pinckus, Kenneth Rexroth and Pitirim Sorokin. For I. F. Stone's views during the Cold War see his *The Truman Era* (New York: Monthly Review Press, 1953), *The Haunted Fifties* (New York: Random House, 1963), and for the early and mid-1960's his *In a Time of Torment* (New York: Random House, 1967).

24. The *Nation* during the Cold War period was edited by Freda Kirchwey (until 1955) and Carey McWilliams (1955–). Its contributors during this period included Louis Adamic, Stringfellow Barr, Carlton Beals, Fred Cook, Julio Alvarez del Vayo, Carl Dreher, D. F. Fleming, Waldo Frank, Mark Gayn, Irving Louis Horowitz, Matthew Josephson, Owen Lattimore, Robert S. Lynd, Archibald MacLeish, Gene Marine, Walter Millis, James R. Newman, Bernard Nossiter, Kenneth Rexroth, Frederick L. Schuman, Edgar Snow, Dan Wakefield, James P. Warburg, Alexander Werth, Raymond Williams, and William Appleman Williams.
25. The *Monthly Review*, edited by Paul Sweezy and Leo Huberman, had such contributors as Paul Baran, W. E. B. DuBois, Albert Einstein, Henry Pratt Fairchild, Joshua Kunitz, Corliss Lamont, Staughton Lynd, Harry Magdoff, F. O. Matthiessen, Carey McWilliams, Scott Nearing, Frederick L. Schuman, I. F. Stone, and William Appleman Williams. Also see Paul Sweezy and Leo Huberman, *The Theory of U.S. Foreign Policy* (New York: Monthly Review Press, 1960). The contributors to *Science and Society* (edited until 1956 by Bernhard J. Stern) included Herbert Aptheker, Paul Baran, Samuel Bernstein, Eugene Genovese, Alvin Gouldner, Maurice Halperin, Irving Louis Horowitz, Corliss Lamont, Otto Nathan, Joseph Starobin, Dirk Struik, and Paul Sweezy. *Masses and Mainstream* (1948–1963) was edited by Samuel Sillen until 1956, Milton Howard (1956–1958), and Charles Humbolt (1958–1960); among its contributors were Herbert Aptheker, W. E. B. DuBois, Howard Fast, Michael Gold, Shirley Graham, V. J. Jerome, A. B. Magil, Joseph North, Joseph Starobin, Anna Louise Strong, and Dirk Struik.
26. On polycentrism see Frank S. Meyer, *The Conservative Mainstream* (New Rochelle, N.Y.: Arlington House, 1969), pp. 308–388, especially pp. 308–327, for his debate with James Burnham; also see the symposium, "Communism Now: Three Views," *Partisan Review*, 23 (Fall, 1956):511–529, for the views of Isaac Deutscher, George Lichtheim (G. L. Arnold), and Irving Howe; G. F. Hudson, "The Paradox of Polycentric Communism," *New Leader*, 40 (January 7, 1957):7–9; *Polycentrism: The New Factor in International Communism*, ed. Walter Laqueur and L. Labedz (New York: Praeger, 1962).
27. Kennan's Reith Lectures were published as *Russia, the Atom and the West* (New York: Harper, 1958). Also see his "Overdue Changes in Our Foreign Policy," *Harper's*, 212 (August, 1956):27–33; an interesting debate on Kennan's disengagement proposals can be found in the *New Republic*, 134 (January–May, 1958).
28. See C. Wright Mills, *The Causes of World War III* (New York: Simon and Schuster, 1958), and especially the reviews in *Dissent*, 6 (Spring, 1959):189–191, by A. J. Muste, and 191–196, by Irving Howe, and the exchange between Mills and Howe, *Dissent*, 6 (Summer, 1959):

295–301; also see H. Stuart Hughes's review article, "A Politics of Peace," *Commentary*, 26 (February, 1959):118–126.

29. See especially the work of Vera Micheles Dean: *The Nature of the Non-Western World* (New York: New American Library, 1957), (ed.) *New Era in the Non-Western World* (Ithaca, New York: Cornell University Press, 1957), and *Builders of Emerging Nations* (New York: Holt, Rinehart and Winston, 1961). Also see Mark Solomon, "Black Critics of Colonialism and the Cold War," in *Cold War Critics,* ed. Paterson, pp. 205–239.
30. For general information see Robert A. Levine, *The Arms Debate* (Cambridge, Mass.: Harvard University Press, 1963). For the battle of the books in the late 1950's and early 1960's see Henry Kissinger, *Nuclear Weapons and Foreign Policy* (New York: Harper, 1957), and his *The Necessity for Choice: Prospects of American Foreign Policy* (New York: Harper, 1960), Thomas C. Schelling, *The Strategy of Conflict* (Cambridge, Mass.: Harvard University Press, 1960), Herman Kahn, *On Thermonuclear War* (Princton, N.J.: Princeton University Press, 1960), and his *Thinking About the Unthinkable* (New York: Horizon Press, 1962), Richard Barnet, *Who Wants Disarmament?* (Boston: Beacon Press, 1960), Norman Cousins, *In Place of Folly* (New York: Harper, 1961), Erich Fromm, *May Man Prevail?* (Garden City, N.Y.: Doubleday, 1961), Seymour Melman, *The Peace Race* (New York: George Braziller, 1962) and H. Stuart Hughes, *An Approach to Peace* (New York: Atheneum Press, 1962). On the formation of SANE in 1957 and on the peace movement in the late 1950's and early 1960's see Wittner, *Rebels Against War,* pp. 240–275, and Nathan Glazer, "The Peace Movement in America—1961," *Commentary,* 31 (April, 1961):288–296.
31. See William Welch, *American Images of Soviet Foreign Policy* (New Haven, Conn.: Yale University Press, 1970), for an indication of how slowly academic-expert thinking on the Soviet Union was changing in the 1960's. Also see the symposium, "The Cold War and the West," *Partisan Review,* 29 (Winter, 1962):9–89, and the symposium, "Liberal Anti-Communism Revisited," *Commentary,* 44 (September, 1967): 31–79.

CHAPTER 5

1. Eric F. Goldman, *The Tragedy of Lyndon Johnson* (New York: Knopf, 1968), p. 513.
2. For an account of the methodology involved, see my article "Reason Analysis," *International Encyclopedia of the Social Sciences,* ed. David Sills, vol. 13 (New York: Macmillan, 1968), pp. 338–343.
3. Nonetheless, 70 percent of respondents named an intellectual who was in turn named by at least one other respondent.
4. The absence of names in our account is not due to mere coyness. We feel that, armed with the names according to frequency of mention, the industrious gossip could figure out who on our chart was talking with whom, a matter which we feel violates confidence.

CHAPTER 6

1. For an outline of radical views, written, however, by someone unsympathetic to them, see Robert W. Tucker, *The Radical Left and American Foreign Policy* (Baltimore: Johns Hopkins, 1971). All but one of our respondents who claimed the Cold War as a cause of the Vietnam War, took the position Tucker identifies with the "liberal-realist critics."
2. Tucker, *The Radical Left*.
3. A shorter version of the interview schedule was used for some telephone interviews and these tended not to go into reasons, but respondents merely stated policy opinions.
4. Note that my use of *ideology* is different from Marx's and carries no pejorative connotation. For an analysis of the many uses of the term and an attempt empirically to dimensionalize it, see Robert D. Putnam, "Studying Elite Political Culture: The Case of Ideology," *American Political Science Review,* 65 (1971):652–683. For a sociological analysis of ideology in terms of its functional role in justifying or criticizing values and norms see Bernard Barber, "Function, Variability and Change in Ideological Systems," in Barber and Alex Inkeles, *Stability and Social Change* (Boston: Little, Brown, 1971). For an historical account of the term see Norman Birnbaum, *The Sociological Study of Ideology, 1940–1960: A Trend Report and Bibliography* (Oxford: Blackwell, 1962).
5. Existentialism, never a dominant mode of American thought, in our terms has neither a systematic method of ascertaining what works nor a preconceived set of ultimates.
6. *Commentary,* 44 (September, 1967):31. Comments in brackets supplied.
7. Alan Silver suggested to me that one subtheme of American intellectual history is a continuous oscillation among intellectuals between pragmatism and idealism.

CHAPTER 7

1. Naturally there is some overlap, since a good many of the leading writers were in our sample. But the distribution of opinion among sample members is not affected when the writers are removed.
2. We have a very accurate representation. Thirteen percent of the 172 sampled leading intellectuals received two or more nominations as having influenced a respondent's opinion on Vietnam. Half of the top 12 were actually interviewed in this study. Seventeen percent of the 172 were named two or more times as having a reputation for influence in the intellectual community; 22 percent were named twice or more as discussion partners and 29 percent received two or more nominations on one or more questions on Vietnam. There are no more than one or two percentage points' difference between the sample as interviewed and the sample as originally drawn on any of these nominations. Thus, our study accurately reflects the opinions of Vietnam influentials.
3. A Tau B. of .25 for grouped data. There is a closer relationship between

discussion of Vietnam and being nominated as a generally influential intellectual – .49; a moderately closer relationship also exists between having a reputation for influence on Vietnam among intellectuals and having a reputation for general influence – .39. As we shall see, when intellectuals think in terms of reputation they tend to think more of better-known intellectuals, so this raises the correlation between lists on different types of issues. Since discussion takes place mainly within the circles of the leading intellectuals, discussion correlates more highly with a general reputation for influence than does any other single indicator of prominence. Naturally, the correlations among our three lists on Vietnam – discussion partners, influence on own thinking and reputation for influence – are quite high: ranking between .60 and .80.

4. Unless qualified by the word "slightly," all relationships reported here show a Kendal's S for ordered data significant past the .05 level, or where appropriate for unordered data, a Chi square of at least .05. These statistics are used merely as rules of thumb or guidelines, since there are, of course, a number of problems in their interpretation for this type of sample.
5. Robert W. Tucker, *The Radical Left and American Foreign Policy* (Baltimore: Johns Hopkins, 1971), p. 12.
6. Ibid., pp. 12–13. Italics ours.
7. Shaplen's later works include much more fundamental criticisms of American policy and verge more toward the moral and ideological, but not in the period when he made his impact on our sample.
8. November 12, 1972, p. 20.
9. In a much less kind review in the *New York Review of Books* (January 25, 1973), Mary McCarthy, who was never noted for her kindness, blasts Halberstam for not taking what we have called an ideological or a moral point of view, thus again reaffirming both the way we classified her earlier works and the way we classify Halberstam.

CHAPTER 8

1. Robert W. Tucker, *The Radical Left and American Foreign Policy* (Baltimore: Johns Hopkins, 1971), p. 1.
2. Question wording is identical to a Gallup poll question. For an analysis of the effect of phrasing on answers to these questions see John E. Mueller, "Trends in Popular Support for the Wars in Korea and Vietnam," *American Political Science Review*, 65 (June, 1971):358–375.
3. Stephen E. Ambrose, *Rise to Globalism* (Baltimore: Penguin Books, 1971), pp. 196–197.
4. Ibid., p. 198.
5. This apt phrase was suggested by Allen Barton.
6. John F. Kennedy, who had attracted around him many of the intellectuals in this sample, had said in that famous speech, so applauded by liberal intellectuals: "Let every nation know, whether it wishes us well or ill, that we shall pay any price, bear any burden, meet any hardship, support any friend, oppose any foes, in order to assure the survival and success of liberty. This much we pledge – and more."
7. Howard Schuman, "Two Sources of Anti-War Sentiment in America,"

American Journal of Sociology, 78 (November, 1972):513–536. Mueller, "Trends in Popular Support," pp. 366–367, argues rather that popular support for *both* the Korean and Vietnam wars declined by 15 percentage points every time the American casualties increased by a factor of 10.

CHAPTER 9

1. By and large the classification is similar to that which will be used for the issue with which they were most concerned. Foreign policy included the war in Vietnam, of course, as well as Cold War issues, and the like. "Culture and values" is similar to what we shall call "culture crisis" but to make it comparable to the views of the rest of the American elite it includes such categories as moral decay and the lack of religious values (hardly the way intellectuals see things!), lack of appreciation and understanding for American institutions, youth and university issues, alienation, polarization, breakdown of authority and so on. Economics includes inflation, unions and labor, income distribution and taxation, problems of technology, unemployment, and the like. Social issues include poverty, welfare, social security, old age, health care, and education. Ecology includes various quality-of-the-environment issues, conservation, and population control.
2. "The Issues of the Sixties: An Exploratory Study in the Dynamics of Public Opinion," *Public Opinion Quarterly*, 37 (Spring, 1973):62–75.
3. The sample of intellectuals we are studying does not have high correlations between most of the social background factors. Nonetheless, because we suspected some interaction between variables, most of the relevant factors were controlled for each other. Thus findings reported are those which held up only after relevant variables were controlled. Interactions are reported in the text. All "findings" are on the order of 20 percentage points difference.
4. Respondents who might have been too young for a given period were told, "If you were too young at the time, please start with that time period in which you first became concerned with such issues." The entire item was given only to respondents who had sufficient time to give us, so that we have responses from only 81 of the 110.
5. The problems are regrouped here in a different order from their actual presentation on the checklist. The titles too are added.

CHAPTER 10

1. Even though the administration saw race as a problem. (See Table 22, Page 226.)

CHAPTER 13

1. By the summer of 1972, the issue had been crystallized by the editors of the *Partisan Review* as "The New Cultural Conservatism," and their

summer issue contains important published opinions on both sides of the matter. The ratio of opinion there seems roughly to correspond with the proportions given here for respondents rather than for the perceived influentials.

CHAPTER 15

1. For example: Karl Mannheim, *Ideology and Utopia* (London: Routledge and Kegan Paul, 1936); Florian Znaniecki, *The Social Role of the Man of Knowledge* (New York: Columbia University Press, 1940); Robert K. Merton, *Social Theory and Social Structure* (New York: Free Press, 1957); Lewis Coser, *Men of Ideas: A Sociologist's View* (New York: Free Press, 1965); *On Intellectuals: Theoretical Case Studies*, ed. Philip Rieff (New York: Doubleday, 1969); Talcott Parsons, "'Intellectual': A Social Role Category," in Rieff, p. 4; J. P. Nettl, "Ideas, Intellectuals, and Structures of Dissent," in Rieff, p. 81; and Edward A. Shils, *The Intellectuals and the Powers and Other Essays* (Chicago: University of Chicago Press, 1972). We cannot pretend here to a review of all of their positions, though some will be taken up in the conclusion of the chapter. Interestingly, our respondents themselves offer just about every shade of opinion and observation contained in these works – most of which have in any case been read by many of the respondents.
2. Just what Yugoslavia really is – Western, Balkan, or what-have-you – stirs constant debate and is hardly a matter we need settle here.
3. While Yugoslav intellectuals can hardly allow themselves to say exactly what they want to, without incurring the wrath of the government, they are much freer to do so than intellectuals in any other society dominated by a governing Communist Party. After all, the leading Yugoslav journal of intellectual opposition, *Praxis*, is subsidized by the government, and though an occasional issue is delayed or impounded, the journal continues as a center for Marxist Humanist thought which is frequently at variance with the official "line."
4. Intellectuals in Yugoslavia were chosen in a way similar to those in the United States, except that because of the smaller size of intellectual circles in a Balkan country, some artists, actors, directors, and film directors tend to belong to the same circles as do the literary intellectuals. Then too, there are "official" intellectual leaders in a socialist country such as heads of academies, the writers' union, and other such organizations. But the large majority of the sample again consisted of persons who wrote a good deal in the right places. The Yugoslav intellectuals were interviewed in the context of a larger study of opinion-making. For published reports in English, see Allen Barton, Bogdan Denitch, and Charles Kadushin, *Opinion-Making Elites in Yugoslavia* (New York: Praeger, 1973).
5. After the checklist just analyzed, we asked: "The relations of intellectuals to men of power (political power) is an issue of concern to many in the intellectual community. What is your thinking about this issue?" Because the question came at the end of a very long interview, not

everyone had a chance to talk about this matter. Seventy-five did; however, almost 60 discussed the "oughts" of the situation, while 50 presented their view of the way things actually were (obviously, 35 did both).

6. Both with respect to the war in Vietnam and with respect to the issue of greatest concern to our respondents, we asked a series of questions such as: "What people do you think are especially important in this country in determining our policy (on the issue, or on Vietnam)?" After this question had been discussed, intellectuals were asked how they would go about getting their ideas across to these "important people" and whether they had any "special channels" to them. Finally, we asked whether an intellectual had actually tried to reach these "important people" and what the outcome of this attempt had been.

   This line of questioning was omitted on short interviews and on telephone interviews. Of the 20 for whom we lack this information there were three or four who we definitely know had such channels. The others probably did not. In the interests of accuracy, however, we are not guessing and are reporting data only as actually given to us by our respondents. This procedure tends to *overestimate* by about 5 percent the proportion who had regular channels. Because intellectuals often lumped together Vietnam and the other issue in discussing their access to government, we too group Vietnam and the issue of choice.
7. Directed by Allen Barton, Bogdan Denitch, Carol Weiss and myself.
8. "The Intellectual as Celebrity," *Dissent,* 20 (Winter, 1973):46–56.
9. Podhoretz, *Making It,* pp. 312–313, italics in original.
10. This judgment was obviously based on their "line," not their political activity.
11. Carol H. Weiss, "Readership of the Leadership," *Public Opinion Quarterly* (Spring, 1974): table 3, shows for all sectors of the leadership the percent reporting *frequent* reading. The percentages drop precipitously. For example, 25 percent of mass media leaders claim to read the *New York Review of Books* "regularly," as compared with 60 percent at least occasionally. Nonetheless, occasional reading is probably sufficient for a trickle-down effect.

CHAPTER 16

1. Florian Znaniecki, *The Social Role of the Man of Knowledge* (New York: Columbia University Press, 1940). Seymour Martin Lipset and Richard Dobson, "The Intellectual as Critic and Rebel: With Special Reference to the United States and the Soviet Union," *Daedalus,* 101 (Summer, 1972):137.
2. Robert K. Merton, "Role of the Intellectual in Public Bureaucracy," *Social Theory and Social Structure,* rev. ed. (New York: Free Press, 1957).
3. Lewis Coser, *Men of Ideas* (New York: Free Press, 1965).
4. J. P. Nettl, "Ideas, Intellectuals and Structures of Dissent," in *On Intellectuals,* ed. Philip Rieff (New York: Doubleday, 1969), p. 86.
5. Talcott Parsons, "The Intellectuals: A Social Role Category," in Rieff,

On Intellectuals, p. 4. E. A. Shils, *The Intellectuals and the Powers and Other Essays* (Chicago: University of Chicago Press, 1972), p. 3. S. N. Eisenstadt, "Intellectuals and Tradition," *Daedalus,* 101 (Spring, 1972):1–20, in an important review, suggests, however, that there is strain in the role of intellectuals vis-à-vis political powers even where intellectuals create within the framework of a tradition. See also Eisenstadt, *The Political Systems of Empires* (New York: Free Press, 1963), pp. 183–197.

6. Karl Mannheim, "The Sociological Problem of the 'Intelligentsia,'" *Ideology and Utopia* (London: Routledge and Kegan Paul, 1936).
7. If our study has any serious gaps, they are most apparent in the area of recruitment. As we noted in Chapter 1, Page 25, at one time we had a series of questions in our pretest dealing with how a respondent became an intellectual. The question is an intriguing one because clearly one does not get one's "job" as an intellectual by answering an advertisement in the *New York Times.* The process is complex, and it took most respondents an hour merely to outline their history. Hence, except for a listing of actual jobs held, the entire section of the interview dealing with recruitment was dropped. We leave this intriguing subject to another research team, and confine our remarks here to a few salient points.
8. Mannheim, "The Sociological Problem of the 'Intelligentsia,'" p. 138.
9. Lipset and Ladd, forthcoming study.
10. Shils, *The Intellectuals and the Powers,* p. 14. For a view that many intellectuals in the past were actually quite isolated from each other, see Richard Hofstadter, *Anti-Intellectualism in American Life* (New York: Knopf, 1963), pp. 425–426, footnote.
11. Diana Crane, *Invisible Colleges: Diffusion of Knowledge in Scientific Communities* (Chicago: University of Chicago Press, 1972), reports a similar finding for the initial stages of a scientific field. Perhaps values and social structure are always mismatched at the initial stage!
12. For a different point of view see Nathan Glazer, "Role of the Intellectuals," *Commentary,* 51 (February, 1971):55–61.
13. Joseph Ben David, "The Growth of the Professions and the Class System," *Current Sociology,* 12 (1963–64):256–277
14. Parsons, "The Intellectual," p. 23.
15. Shils, *The Intellectuals,* pp. 13, 154–155.
16. Nettl, in Rieff, p. 81.
17. Ibid., p. 86; Coser, *Men of Ideas,* p. x.
18. Shils, *The Intellectuals,* p. 154.
19. Exceptions are Coser's series of essays in *Men of Ideas* and Wrong's review of the *New York Review of Books* ("The Case of the *New York Review,*" *Commentary,* 50 [November, 1970]:49–63), but that review was written in Wrong's role as an intellectual, not as a professional sociologist. Sociologists of *science,* a field closely akin to the sociology of intellectuals and indeed sometimes considered part of it, have, in contrast, found the dissemination and publication of ideas to be one of the most important areas for research. Historians, journalists, and intellectuals themselves have had a great interest in intellectual journals in the last few years as a result of some apparent changes of the sixties. See, for example, Peter Steinfels, "The Cooling of the Intellectuals:

The Case of *Commentary* and the *New York Review of Books*," *Commonweal*, 94 (May 21, 1971):255–261; Merle Miller, "Why Norman and Jason Aren't Talking," *New York Times Magazine* (March 26, 1972); Philip Nobile, "A Review of the *New York Review of Books*," *Esquire*, 77 (April, 1972); Richard Kostelanetz, "Literary Power and Literary Violence," *Unmuzzled Ox*, 1 (February, 1972); Irving Howe, "The New York Intellectuals: A Chronicle and a Critique," *Commentary*, 46 (1968):29–51. A substantial work is James B. Gilbert, *Writers and Partisans: A History of Literary Radicalism in America* (New York: John Wiley, 1968).

20. Richard Hofstadter, *Anti-Intellectualism in American Life*, p. 401.
21. Ibid., p. 406.
22. Ibid., p. 408.
23. Ibid., p. 410.
24. Ibid., p. 412.
25. Ibid., p. 394.
26. Ibid., p. 415.
27. *Commentary*, 1968.
28. Howe, "The New York Intellectuals," p. 37.
29. Ibid.
30. Ibid., pp. 37–38.
31. Ibid., p. 33.
32. Seymour Martin Lipset and Richard B. Dobson, "The Intellectual as Critic and Rebel," pp. 137–198; Jill Conway, "Intellectuals in America: Varieties of Accommodation and Conflict," *Daedalus*, 101 (Summer, 1972):199–205; Martin E. Malia, "The Intellectuals: Adversaries or Clerisy?" *Daedalus*, 101 (Summer, 1972):206–216.
33. David Kettler, "The Vocation of Radical Intellectuals," *Politics and Society*, 1 (November, 1970):23–49. Kettler advances the interesting hypothesis that the abandonment of rigorous theorizing and the embracement of immediate experience by the young radicals of the sixties was caused by the change in scope of what could be considered "scientific." In Marx's day, he points out, it was quite legitimate to believe that values and prescriptions could be deduced "scientifically" from social and historical analysis. The realization by twentieth-century epistemology that such deductions were not scientifically possible threw the entire enterprise of rigorous leftist social theorizing into doubt. One reaction to this doubt was the abandonment of logic in favor of experience. Kettler of course supports the idea that much in the way of valid Left theorizing can be accomplished within the framework of contemporary social science, given that the theorizer has a basic commitment to a radical point of view. Nettl, "Ideas, Intellectuals, and Structures of Dissent," p. 102, also notes the problems made by science for the purveyors of radical values.
34. Norman Podhoretz has taken a public beating among many intellectuals for having himself written in the late 1960's, or had others write in *Commentary*, what most intellectuals by 1970 had come strongly to feel.
35. Howe, "The New York Intellectuals," p. 44.
36. For a sophisticated view of this process see Amitai Etzioni, *The Active Society* (New York: Free Press, 1968), pp. 182–189.

37. Though in his recent *The American University,* Parsons uses "intellectual" as I do.

APPENDIX

1. Charles Kadushin, Julie Hover, and Monique Tichy, "How and Where to Find the Intellectual Elite in the United States," *Public Opinion Quarterly,* 35 (Spring, 1971):1–18.
2. Each person, on the average, appeared in 1.8 sample strata.
3. Editors included 23 persons who were chief editors of the journals mentioned above, with the exception of *Reporter,* which had ceased publication.

Selected Bibliography

AMBROSE, STEPHEN E. *Rise to Globalism*. Baltimore: Penguin Books, 1971.
BARBER, BERNARD. "Function, Variability and Change in Ideological Systems," in Bernard Barber and Alex Inkeles, *Stability and Social Change*. Boston: Little, Brown, 1971.
BARTON, ALLEN H., BOGDAN DENITCH, and CHARLES KADUSHIN (eds.), *Opinion-Making Elites in Yugoslavia*. New York: Praeger, 1973.
BIRNBAUM, NORMAN. *Sociological Study of Ideology 1940–1960: A Trend Report and Bibliography*. Oxford: Blackwell, 1962.
BELL, DANIEL. "Comments on American Intellectuals." *Daedalus*, Summer, 1959.
BEN DAVID, JOSEPH. "The Growth of the Professions and the Class System," *Current Sociology*, 12 (1963–64):256–277.
BEN DAVID, JOSEPH, and AWRAM ZLOCZOWER. "Universities and Academic Systems in Modern Societies." *Archives européennes de Sociologie*, 3 (1962):45–85.
BON, FRÉDÉRIC, and MICHEL-ANTOINE BURNIER. *Les Nouveaux Intellectuels*. Paris: Editions Cujas, 1966.
BREASTED, MARY. "What New York Does to Its Writers." *Village Voice*, June 1, 1972.
BROWN, FRANCIS (ed.). *Opinions and Perspectives*. New York: Peregrine, 1964.
CLARK, TERRY, and PRISCILLA P. CLARK. "Le Patron et son cercle: Clef de l'université française," *Revue Française de Sociologie*, 12 (1971):19–39.
CONFINO, MICHAEL. "On Intellectuals and Intellectual Traditions in Eighteenth and Nineteenth Century Russia." *Daedalus*, 101 (Spring, 1972): 117–149.
CONWAY, JILL. "Intellectuals in America: Varieties of Accommodation and Conflict." *Daedalus*, 101 (Summer, 1972):199–205.
COSER, LEWIS. *Men of Ideas: A Sociologist's View*. New York: Free Press, 1965.
———. "The Intellectual as Celebrity." *Dissent*, 20 (Winter, 1973), pp. 46–56.
CRANE, DIANA. *Invisible Colleges: Diffusion of Knowledge in Scientific Communities*. Chicago: University of Chicago Press, 1972.
EISENSTADT, S. N. "Intellectuals and Tradition." *Daedalus*, 101 (Spring, 1972):1–20.
ETZIONI, AMITAI. *The Active Society*. New York: The Free Press, 1968.
FUNKHOUSER, G. RAY. "The Issues of the Sixties: An Exploratory Study in the Dynamics of Public Opinion." *Public Opinion Quarterly*, 37 (Spring, 1973):62–75.
FYVEL, T. R. *Intellectuals Today: Problems in a Changing Society*. New York: Schocken, 1968.
GILBERT, JAMES B. *Writers and Partisans: A History of Literary Radicalism in America*. New York: Wiley, 1968.
GLAZER, NATHAN. "Blacks, Jews and the Intellectuals." *Commentary*, 47 (April, 1969), pp. 33–39.
———. "Role of Intellectuals." *Commentary*, Vol. 51 (February, 1971).
HARRINGTON, MICHAEL. "We Few, We Happy Few, We Bohemians." *Esquire* (August, 1972), pp. 99–103, 162–164.
HOFSTADTER, RICHARD. *Anti-Intellectualism in American Life*. New York: Knopf, 1963.

HOWE, IRVING. "The New York Intellectuals: A Chronicle and a Critique." *Commentary,* 46 (October, 1968):29–51.

KADUSHIN, CHARLES. "The Friends and Supporters of Psychotherapy on Social Circles in Urban Life." *American Sociological Review,* 31 (December, 1966), 786–802.

———. "Power, Influence and Social Circles: A New Methodology for Studying Opinion-Makers." *American Sociological Review,* 33 (1968): 685–699.

———. "Reason Analysis," in *International Encyclopedia of the Social Sciences,* ed. Davis Sills. New York: Macmillan, 1968, XII, 338–343.

KADUSHIN, CHARLES, JULIE HOVER, and MONIQUE TICHY. "How and Where to Find Intellectual Elite in the United States." *Public Opinion Quarterly,* 35 (Spring, 1971): 1–18.

KADUSHIN, MAX. *The Rabbinic Mind.* New York: Blaisdell, 1965.

KETTLER, DAVID. "The Vocation of Radical Intellectuals," *Politics and Society,* 1 (November, 1970):23–49.

KOSTELANETZ, RICHARD. "Literary Power and Literary Violence," *Unmuzzled Ox,* 1 (February, 1972).

LADD, EVERETT CARL, JR. "Professors and Political Petitions," *Science,* 163 (March 28, 1969):1425–1430.

LADD, EVERETT CARL, JR., and SEYMOUR MARTIN LIPSET. "Politics of Academic Natural Scientists and Engineers," *Science,* 176 (June 9, 1972): 1091–1100.

LASCH, CHRISTOPHER. *The New Radicalism in America, 1889–1963: The Intellectual as a Social Type.* New York: Knopf, 1965.

———. *The Agony of the American Left.* New York: Knopf, 1968.

LINZ, JUAN. "Intellectual Roles in Sixteenth and Seventeenth Century Spain," *Daedalus,* 101 (Summer, 1972):59–108.

LIPSET, SEYMOUR MARTIN. *Political Man.* New York: Doubleday, 1960.

LIPSET, SEYMOUR MARTIN, and EVERETT CARL LADD. "Jewish Academics in the United States," *American Jewish Yearbook,* 72 (1971):89–128.

LIPSET, SEYMOUR MARTIN, and RICHARD DOBSON. "The Intellectual as Critic and Rebel: With Special Reference to the United States and the Soviet Union," *Daedalus,* 101 (Summer, 1972):137–198.

MACDONALD, DWIGHT. "Norman Cousins' Midcult (World)," *Columbia Forum,* 1 (Fall, 1972):18–25.

MALIA, MARTIN E. "The Intellectuals: Adversaries or Clerisy?" *Daedalus,* 101 (Summer, 1972):206–216.

MANNHEIM, KARL. *Ideology and Utopia.* London: Routledge and Kegan Paul, 1936.

MERTON, ROBERT K. *Social Theory and Social Structure.* New York: Free Press, 1957.

———. "Insiders and Outsiders: A Chapter in the Sociology of Knowledge," *American Journal of Sociology,* 78 (July, 1972):9–47.

MILGRAM, STANLEY. "The Small-World Problem," *Psychology Today,* 1 (1967).

MILLER, MERLE. "Why Norman and Jason Aren't Talking," *New York Times Magazine* (March 26, 1972).

MUELLER, JOHN E. "Trends in Popular Support for the Wars in Korea and Vietnam," *American Political Science Review,* 65 (June, 1971):358–375.

NETTL, J. P. "Ideas, Intellectuals, and Structures of Dissent" in *On Intel-*

lectuals: Theoretical Studies, Case Studies, ed. Philip Rieff. New York: Doubleday, 1969.
NOBILE, PHILIP. "A Review of the *New York Review of Books," Esquire,* Vol. 77 (April, 1972).
PARSONS, TALCOTT. " 'The Intellectual': A Social Role Category," in *On Intellectuals: Theoretical Case Studies,* ed. Philip Rieff. New York: Doubleday, 1969.
PARSONS, TALCOTT, and GERALD M. PLATT. *The American University.* Cambridge, Mass.: Harvard University Press, 1973.
PODHORETZ, NORMAN. *Making It.* New York: Random House, 1967.
PUTNAM, ROBERT D. "Studying Elite Political Culture: The Case of Ideology," *American Political Science Review,* 65 (1971):652–683.
RIEFF, PHILIP (ed.). *On Intellectuals: Theoretical Case Studies.* New York: Doubleday, 1969.
SCHUMAN, HOWARD. "Two Sources of Anti-War Sentiment in America," *American Journal of Sociology,* 78 (November, 1972):513–536.
SHILS, EDWARD A. *The Intellectuals and the Powers and Other Essays.* Chicago: University of Chicago Press, 1972.
SNOW, C. P. *The Two Cultures and the Scientific Revolution.* New York: Cambridge University Press, 1959.
STEINFELS, PETER. "The Cooling of the Intellectuals: The Case of *Commentary* and *The New York Review of Books," Commonweal* (May 21, 1971):255–261.
STONE, I. F. *The Hidden History of the Korean War.* New York: Monthly Review Press, 1952.
Times Literary Supplement. "The Art of Science," October 25, 1963. Articles by C. P. Snow, P. B. Medawar, others.
TRAVERS, JEFFREY, and STANLEY MILGRAM. "An Experimental Study of the Small-World Problem," *Sociometry,* 32 (December, 1969):425–443.
TUCKER, ROBERT W. *The Radical Left and American Foreign Policy.* Baltimore, Johns Hopkins, 1971.
WELCH, WILLIAM. *American Images of Soviet Foreign Policy.* New Haven: Yale University Press, 1970.
WHITE, CYNTHIA and HARRISON. *Canvases and Careers: Institutional Change in the French Painting World.* New York: Wiley, 1965.
WRONG, DENNIS H. "The Case of the *New York Review," Commentary,* 50 (November, 1970):49–63.
ZNANIECKI, FLORIAN. *The Social Role of the Man of Knowledge.* New York: Columbia University Press, 1940.
ZUCKERMAN, HARRIET. "Stratification in American Science," *Sociological Inquiry,* 40 (Spring, 1970):235–257.
———. *Scientific Elites: Studies of Nobel Laureates in the United States.* Chicago: University of Chicago Press, 1973.
ZUCKERMAN, HARRIET, and ROBERT K. MERTON. "Patterns of Evaluation in Science: Institutionalization, Structure and Functions of the Referee System," *Minerva,* 9 (January, 1971):66–100.